Chemistry 332L

Essentials of Organic Chemistry Laboratory

UNIVERSITY
OF
SOUTH CAROLINA

Department of Chemistry and Biochemistry

**QDE PRESS
2015**

Copyright 2015 © by George Handy and Salman Ali
Copyright 2015 © Illustrations by Salman Ali and Byron Farnum
Copyright 2015 © QDE Press Inc.

All rights reserved.

Permission in writing must be obtained from the publisher before any part of this work may be reproduced or transmitted in any form or by any means, electronic or mechanical, including photocopying and recording, or by any information storage or retrieval system.

Printed in the United States of America.

ISBN: 978-1-938535-09-3

QDE Press
8828 Autumnbrooke Way
Montgomery, Al 36117
www.qdepress.com

Acknowledgements:

Salman Ali
Jerry Oxsher
George Handy
Mitul Patel

Revised and edited by the following:

Pranv Patel
Jesus Perez
Jeffrey Steen
Hal Addison

Table of Contents

Introduction	135
Waivers	136
Introduction to ^{1}H-NMR	145
Experiment 1: Identification of an Unknown Aldehyde or Ketone	181
Experiment 2: A Grignard Reaction	203
Experiment 3: An Aldol Condensation	219
Experiment 4: Sodium Borohydride Reduction	233
Experiment 5: Acid-Catalyzed Esterification	247
Experiment 6: Diels-Alder Reaction	263
Experiment 7: Solvent-Free Claisen Condensation Of Ethyl Phenylacetate	273
Experiment 8: EAS: Nitration of Methyl Benzoate	287
Experiment 9: EAS: Friedel-Crafts Alkylation	303
Experiment 10: Click Reaction	317
Appendices	329

Organic Chemistry Laboratory

Laboratory Coordinator: Dr. George Handy
Office: PSC 218A
Office hours: Laboratory hours

Textbook: This manual is the only text required for this course. All notes should be written down *in this manual.* No extra notebook is required.

Check-in/out: You must check a list of equipment in your drawer at the beginning of the semester. Any missing or broken glassware must be replaced by your TA. Each week, your TA will check the contents of your drawer before you are permitted to leave. You must check out your drawer with your TA at the end of the semester or your grades will not be released.

Safety: In order to ensure the safety of everyone in the lab, several waivers must be signed and turned in to the TA <u>before</u> you are permitted into the lab. All prerequisite and safety waivers are located in this manual.

Grading: Please See Syllabus

- <u>Technique</u>: Points will be deducted from your lab report if you are messy or unsafe.

Your TA will notify you if any changes are made to this format or to the syllabus.

Please fill out the following forms and leave them in your lab manual for your TA to check at any time, or in the case of an accident.

Chemistry 332L
Prerequisite

Name of student (please print):

Major:

Note:

In order to continue into Chem 334L, all students must have completed Chem 333L, Chem 333 and Chem 334. There is an exception for those who are currently enrolled in Chem 334.

Chem 333L information

1. Year, semester, and institute:

2. Grade received:

3. Instructor:

Chem 333 information

1. Year, semester, and institute:

2. Grade received:

3. Instructor:

Chem 334 information for those who have completed the class.

1. Year, semester, and institute:

2. Grade received:

3. Instructor:

Chem 334 information for those who are currently enrolled in the class.

1. Instructor:

Signature: _____

UNIVERSITY OF SOUTH CAROLINA
Personal Safety in the Undergraduate Organic Lab

It is a policy of the University of South Carolina Department of Chemistry and Biochemistry's Division of Undergraduate Organic Labs to require certain personal safety equipment. This includes approved eye protection. All students are to wear approved Z87-type goggles (not safety glasses) at all times when in the lab. There are to be no exceptions to this rule. Your lab TA will send you home either to get approved goggles or make up the lab at another time.

In addition, all students are to wear closed-toed shoes (no sandals). You will be required to leave the lab if you are not wearing proper shoes. Again, there are no exceptions to this rule.

It is recommended by the Department of Chemistry and Biochemistry's Division of Undergraduate Organic Labs that students wear long pants (no shorts) and shirts with sleeves. (Natural fabrics are always better.) However, it is the individual student's decision as to whether shorts (as opposed to pants) are worn. It is in this regard that the individual students are to work at their own risk. Understanding and researching the hazards associated with the chemicals and equipment used is a pre-laboratory exercise for all labs. It is advised that the student wear suitable clothing for a given laboratory exercise.

I understand and agree to abide by the policy regarding personal safety equipment (see above) in the undergraduate organic chemistry laboratory.

Name (Printed): _____

Student Number #: _____

Course: _____

Lab TA: _____

Lab Section: _____

Signature: _____ Date: _____

DEPARTMENT OF CHEMISTRY

UNIVERSITY OF SOUTH CAROLINA
SAFETY POLICIES

Safe practice in the chemical laboratory is a mutual responsibility and requires the full cooperation of everyone concerned at all times. This cooperation means that each student and instructor will observe safety precautions and procedures. The following general safety rules will be rigidly and impartially enforced throughout the semester. Noncompliance may result in dismissal from the lab (with the time missed to be made up on your own) and/or may result in a grading penalty.

1. Safety goggles must be worn at all times anywhere in the laboratory, even when not performing an experiment. Contact lenses should not be worn during the lab period.

2. Footwear should provide adequate protection against possible safety hazards (broken glass, reagent, etc.).

3. Food or drink will not be allowed in the laboratory. Smoking is also not permitted.

4. Horseplay or other acts of carelessness are prohibited.

5. Unauthorized experiments are not permitted. Unapproved variations in experiments, including changing the quantities of reagents, may be dangerous.

6. Every student is responsible for keeping his work area neat and orderly. Clean the equipment and store away correctly before you leave.

7. The instructor should be informed immediately of any safety hazards or accidents.

All accidents have causes and therefore can be prevented. Many accidents result from a lackadaisical attitude and from a failure to use common sense and to follow instructions. Be aware of what your neighbors do – you may be a victim of their accidents. Do not hesitate to comment tactfully to a neighbor whom you observe engaging in an unsafe practice. Thoroughly acquaint yourself with the location and use of emergency equipment (fire extinguishers, eye-wash stations, showers, etc.) around the lab. With the positive approach of good safety practice, all personal injuries can be avoided.

I have read and understood the safety rules outlined above. I agree to abide by them at all times while participating in Chemistry _____ laboratory, Section_____.

(signed)_____

(date)_____

Teaching Assistant Name: _____

Laboratory meeting day and time: _____

Laboratory room number: _____

Laboratory Safety Sheet

In addition to the safety policies outlined in the syllabus, you are required to read and understand the following policies regarding behavior, work habits, personal safety (especially eyes and hands).

1. You must wear safety goggles / glasses at all times. You will be asked to leave if you are found not wearing your goggles.
2. You must wear closed toe / heel shoes to lab. You will be asked to leave if you are found not wearing appropriate footwear.
3. Eating, drinking or smoking is not permitted in lab.
4. Horseplay or other acts of carelessness are prohibited.
5. Unauthorized experiments are not permitted. Unapproved variations in experiments, including changes in the quantity of reagents, are dangerous.
6. Keep your work area clean during experiments. NO BOOKBAGS, CLOTHING OR UMBRELLAS ARE ALLOWED ON THE FLOOR. Also, you will not be permitted to leave until your work area has been inspected by the TA.
7. Notify your instructor immediately in the case of any safety hazards or accidents.

Other recommendations:

1. It is recommended that you wear long pants and shirts with long sleeves. Also, synthetic fabrics are more prone to catch fire or degrade in the presence of certain reagents. If you choose to not follow these recommendations, you do so at your own risk.
2. Contact lenses increase the risk of eye damage in. the case of an accident.
3. Persons working in laboratories without wearing safety gloves can have an increase in injury from chemical splashes. It is recommended that Nitrile safety gloves be worn in laboratories in which chemicals are handled.

Chemicals splashed on the skin can be absorbed which can cause irritations, burns and possibly affect internal organs. The Nitrile gloves will keep the majority of chemicals from coming into contact with your skin. Washing your hands after contact with chemicals does not ensure that the chemicals did not enter your body.

In the case that chemicals do penetrate through the gloves, immediately remove the gloves, wash your hands and use a fresh pair of gloves.

I have read and understand the safety policies and warnings above.
I understand the risks of wearing contact lenses.
I understand the risks of not wearing gloves.

Signed: _____   Date: _____

CONTACT LENSES IN CHEMICAL LABORATORIES

The following was taken from "Handbook of Laboratory Safety," N. V. Steele, Ed., 2nd ed., Chemical Rubber Co., Cleveland, OH, 1971:

"Contact lenses worn by persons working in laboratories can increase injury from chemical splashes because the wearer may not be able to remove the lenses to permit thorough irrigation, and a person giving first aid may not know that contact lenses are being worn or how to remove them. It is recommended that contact lenses not be worn in laboratories in which chemicals are handled or that wearers be sure to use full eye protection at all times."

In the pamphlet "Use of Contact Lenses in Industry," published by the Council of Occupational Health of the American Medical Association, there are three paragraphs, which are particularly applicable to wearing of contact lenses in laboratories:

"Many physicians believe that the substitution of contact lenses for spectacles in industrial workers is contraindicated in workers whose eyes may be exposed to dust, molten metals, or irritant chemicals. Small foreign bodies, which normally are washed away by tears, sometimes become lodged beneath contact lenses, where they may cause injury to the cornea. Similarly, chemicals splashed into the eye may be trapped under a contact lens can cause extensive damage before the lens can be removed and the eye adequately irrigated."

"For effective protection for the eyes, the contact lens wearer should use the same approved face shields, conventional safety spectacles, or goggles for protection against job hazards as would any other worker on a similar job. Since removal of a contact lens for urgent irrigations after injury is made so difficult by spasm of the eyelids, the contact lens wearer is in even greater need of these protections than their counterpart who does not wear contact lenses if the job carries high potential risk of eye injury."

"Contact lenses are not in themselves protective devices and in fact may increase the degree of injury to the eyes. The same eye-protective devices used by other workers should be worn by contact lens wearers in similar employment."

I, _____, (print) have been informed of and have understood the hazards associated with wearing contact lenses in the chemistry laboratory. I agree to wear safety goggles at all times while participating in Chemistry _____ Laboratory.

Signed: _____

Date: _____

Sect. #: _____

NITRILE GLOVES IN CHEMICAL LABORATORIES

Persons working in laboratories without wearing safety gloves can have an increase in injury from chemical splashes. It is recommended that Nitrile safety gloves be worn in laboratories in which chemicals are handled.

Chemicals splashed on the skin can be absorbed, which can cause irritations, burns, and possibly can affect internal organs. The Nitrile gloves will keep the majority of organic compounds from reaching the skin. Many organic compounds will penetrate through the skin and travel through the body, which can possibly cause extensive damage. Washing your hands immediately does not insure that the chemical(s) did not enter your body. For effective protection for the hands, the person performing the lab experiment should wear Nitrile approved safety gloves. Latex gloves have been found to be very ineffective when working with organics. Most organics can penetrate through these gloves.

Nitrile safety gloves are not full proof protective devices. Some organic compounds can penetrate through these gloves, but the majority of the chemicals used in these labs will not penetrate through them. These Nitrile gloves also have the potential to increase the degree of injury to the hands. This could be due to the organics that can penetrate through the gloves are now held even closer to the skin, while the gloves are still on. It is imperative that if the chemical(s) penetrates through the gloves that the gloves are immediately removed and the person's hands are washed thoroughly under running water and then a fresh pair of gloves be replaced. There is also the issue that gloves can cause a person to lose some of their dexterity. This can also be a potential risk to the individual handling glassware or chemicals. Other workers in similar employment should use the same hand-protective devices.

I, _____, (print) have been informed of and have understood the hazards associated with not wearing Nitrile safety gloves in the chemistry laboratory.

Signed: _____

Date: _____

Course: _____

Sect. # _____

University of South Carolina Department of Chemistry & Biochemistry

CHEMISTRY 332L INVENTORY SHEET

Name: _____
TA: _____
Drawer #: _____

ITEM	QUANTITY IN	QUANTITY OUT
Round Bottom Flask (25 or 50 mL)	_____	_____
Side Arm Adapter (Distillation)	_____	_____
Thermometer	_____	_____
Thermometer Adapter	_____	_____
Condenser/Distilling Column	_____	_____
Graduated Cylinder (10 mL)	_____	_____
Graduated Cylinder (25 mL)	_____	_____
Beaker (50 mL)	_____	_____
Beaker (100 or 150 mL)	_____	_____
Beaker (250 mL)	_____	_____
Beaker (400 mL)	_____	_____
Separatory Funnel (250 mL)	_____	_____
Separatory Funnel Stopper	_____	_____
Buchner Funnel	_____	_____
Wide Mouth Short Stem Funnel	_____	_____
Spatula	_____	_____
Scoopula	_____	_____
Small Test Tubes (6)	_____	_____
Large Test Tubes (6)	_____	_____
Test Tube Rack	_____	_____
Test Tube Holder	_____	_____
Glass Stirring Rod	_____	_____
Erlenmeyer Flask (25 mL)	_____	_____
Erlenmeyer Flask (50 mL)	_____	_____
Erlenmeyer Flask (125 mL)	_____	_____
Erlenmeyer Flask (250 mL)	_____	_____
Filter Flask (250 or 125 mL)	_____	_____
Suction Filtration Adapter	_____	_____

Student Signature: _____

TA Signature: _____

MAP OF THE LAB ROOMS AND SAFETY EQUIPMENT

In the space provided below, draw a diagram of your laboratory, labeling all exits, your workstation's location, and safety equipment (hoods, body showers, eye washes, fire extinguishers, etc).

CHEM 332L

Intro to ¹H-NMR

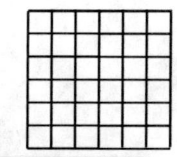

¹H-NMR Spectroscopy

Instructions:

A. Review the material on the theory of NMR and interpretation of NMR.

B. Answer the review questions given.

C. Complete a problem set that you will receive from the TA.

D. Review the material on IR from 333L.

E. A quiz will be given over ¹H-NMR and IR Spectroscopy.

Note:

- You may work in groups on the worksheets.
- The Worksheet will be worth 50 points.
- The spectroscopy quiz will be worth 50 points.

Applications of NMR

a) <u>For chemists:</u>
   i) Assigning the spectra of complex organic molecules
   ii) Detecting weak signals (products of reactions, damaged DNA, etc.)
   iii) Looking at other nuclei (valuable for inorganic chemists for measuring diffusion for chemical warfare or desalination of sea water)
   iv) Studying solids, including plastics or polymers
      (1) Orientation (polarized sunglasses, LCD displays)
      (2) Dynamics (polycarbonates: safety glasses, bulletproof glass)

b) <u>For biologists:</u>
   i) Looking at metabolites in vivo (for example, the rate of ATP hydrolysis)
   ii) Determining how bio-molecules interact (drug binding, protein folding, receptor binding)

c) <u>For pharmaceutical industry:</u>
   i) Determining structure and conformation of products (for example, by mimicking difficult-to-synthesize natural products with synthetic molecules containing the right arrangement of functional groups)

d) <u>For medical profession:</u>
   i) Non-invasively obtaining images of normal, living patients (MRI's of injured athletes, tumors, swelling, internal bleeding)
   ii) Introducing enriched or otherwise contrast-inducing chemicals into the body and following them (for example, monitoring a labeled drug to see if it selectively goes to the target site)

Nuclear Magnetic Resonance
OUTLINE

1. Theory

   A. Introduction

   B. Magnetic Properties of Nuclei

   C. Absorption of Energy

   D. Sheilding/Desheilding

   E. Chemical Equivalence

   F. Spin-Spin Interaction

   G. The Coupling Constant

2. Interpretation of ^{1}H-NMR

   A. Chemical Shifts

   B. Integration

   C. Splitting

3. Introduction to ^{13}C-NMR

1. Theory

A. Introduction to ¹H-NMR

What does ¹H-NMR tell you about a compound??

EX:

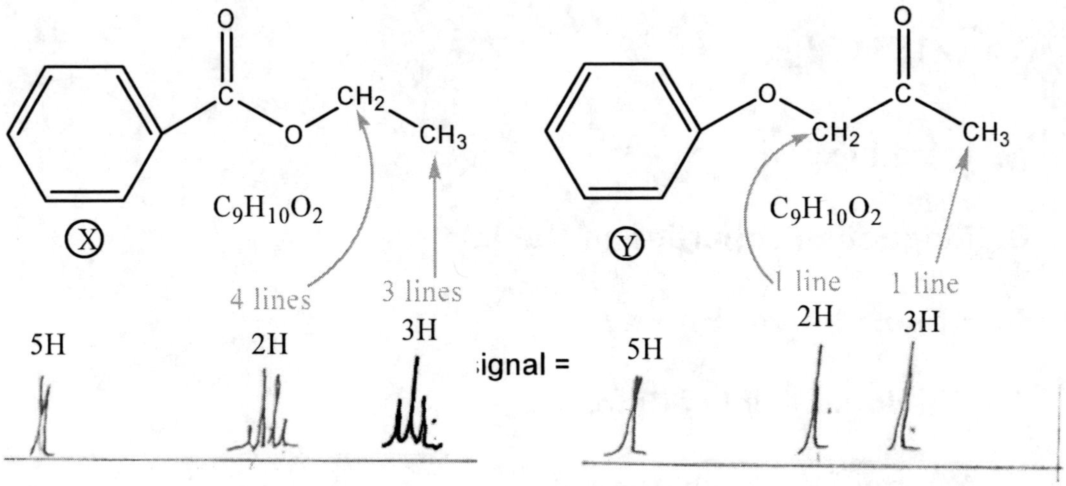

$C_9H_{10}O_2$ (X): Ph–C(=O)–O–CH₂–CH₃
- 5H (benzene)
- 2H, 4 lines (CH₂)
- 3H, 3 lines (CH₃)

$C_9H_{10}O_2$ (Y): Ph–O–CH₂–C(=O)–CH₃
- 5H, 1 line
- 2H, 1 line
- 3H, 1 line

signal =

2. Individual signals = different types of hydrogens.

X has 3 types
Y has 3 types

3. Numbers above each signal represents the number of protons of that specific type that is responsible for that signal.

3 types
- X has 5 of one type
- 2 of one type
- 3 of one type

3 types
- Y has 5 of one type
- 2 of one type
- 3 of one type

4. Signal position gives the chemical environment of the protons.

5. Multiplicity of a signal gives the number of protons that are on the adjacent carbon that are non-equivalent.

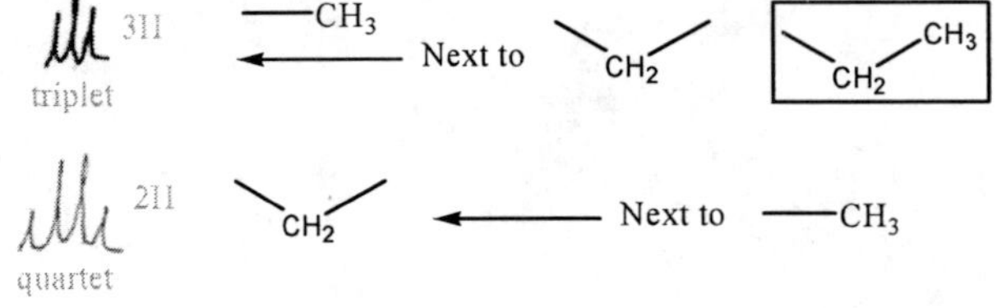

- triplet (3:1:1) ← —CH₃ Next to \CH₂/   \CH₃/CH₂
- quartet (2:1:1) \CH₂/ ← Next to —CH₃

B. Magnetic Properties of Nuclei

Γ) Nuclear Magnetic Moment: μ

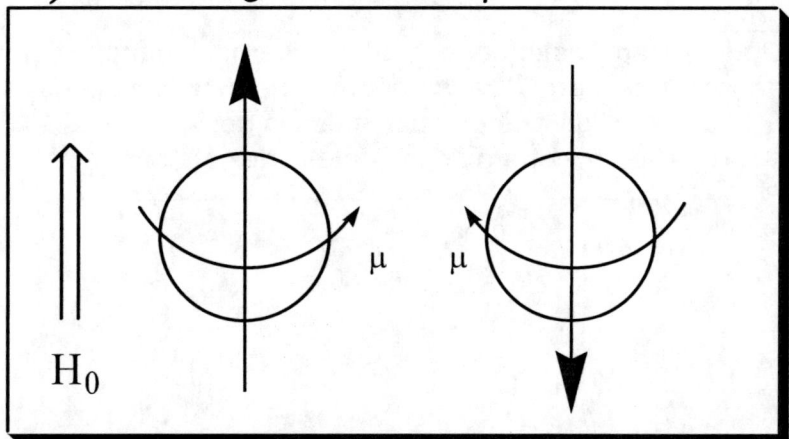

i.) Ho = Applied Magnetic Field
ii.) μ = Induced Magnetic Field
  (1) Created by the circulation of nuclear charge.

2. Angular momentum
  i.) The angular momentum of the spinning nucleus is described in terms of the spin number I of that particular nucleus.
   (1) I can have values of 0, ½, 1, $^3/_2$, etc . . .

3. Spin Number I
  i.) The numerical value of the spin number I is based on the mass number /atomic number combination.
  ii.) Example:

(1) Active in NMR
  Mass 3 = odd
  Atomic # = odd or even
  I = ½ $^3/_2$, etc ...
  $^{13}C_6$ $^1H_1$ $^{31}P_{15}$

(2) Inactive
  Mass # = even
  Atomic # = even
  I = 0
  $^{12}C_6$ $^{16}O_8$

(3) Active in NMR
  Mass # = even
  Atomic # = odd
  I = 1, 2, etc...
  $^2H_1$ $^{14}N_7$

4. Spin States
  i) Nuclei with a finite spin number can exist in spin states according to the expression: 2I + 1.

Example:
(1) In the absence of an external magnetic field, both spin states of a proton have equal energy. <u>They are called Degenerate Energy Levels</u>.
(2) Upon the application of a uniform external magnetic field the degenerate energy levels become unequal spin states.
    (a) The High energy spin states is designated as $-1/2$
    (b) The Low energy spin state is designated as $+1/2$

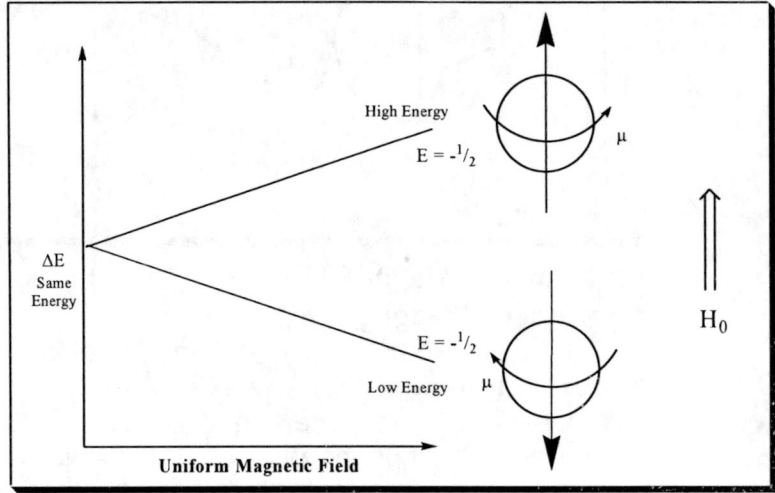

C. Absorption of Energy

a) Energy of the $-1/2$ spin state = $E_2 = \mu H_0$ (High Energy)
b) Energy of the $+1/2$ spin state = $E_1 = -\mu H_0$ (Low Energy)
c) Energy difference between the two unequal spin states.
    1.) $\Delta E = E_2 - E_1$
            $= \mu H_0 - \mu H_0$
            $= 2 \mu H_0$
    2.) $\Delta E = h\nu$ (energy of transition from low to high at resonance)
    3.) $h\nu = 2\mu H_0$
            H_0 = Homogeneous Magnetic Field
            H = Planck's Constant
            ν = Radio Frequency
            μ = Nuclear Magnetic Moment

TERMS:

1. <u>Chemical Shifts of protons:</u> Reported in delta values (δ), numerical values indicating the position of absorption of hydrogens with reference to an internal standard, usually TMS set at 0 ppm.

2. <u>Downfield Absorption:</u> Deshielded protons (require less energy). To transmit from $+^1/_2 \Rightarrow -^1/_2$ spin state (seen to the left of the spectrum.

3. <u>Upfield Absorption:</u> Shielded protons (requires more energy). To transmit from $+^1/_2 \Rightarrow -^1/_2$ spin state (seen to the right of the spectrum).

4. <u>Nonequivalent protons:</u> Protons are of different energy and give different chemical shifts due to their chemical environment.

5. <u>Equivalent Protons:</u> Protons that are in identical chemical environments.

Chemical Shift Parameter: HOW TO DETERMINE SIGNAL POSITIONS USING A STANDARD

a) Upfield / Downfield Absorption is relative in order to indicate the signal positions in a more quantitative fashion. A reference standard TMS (tetramethyl silane), in internal standard, is used.

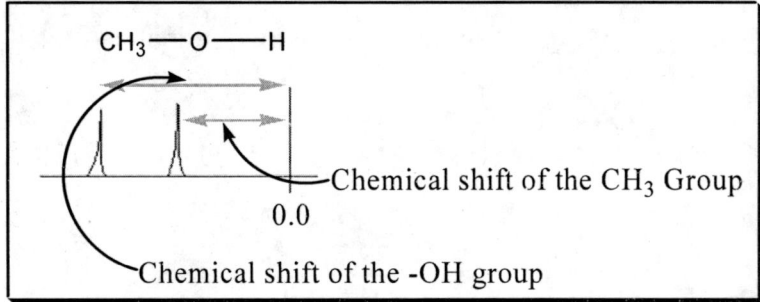

Silane

$$\delta = \frac{\text{Shift in Frequenncy from TMS (Hz)}}{\text{Frequency of Spectrometer (Hz)}} \times 10^6$$

b) The difference between the positions of absorption of TMS and that of a particular proton or a group of protons is called the chemical shift.

c) The chemical shift parameter is given in Delta values.

CH_3—O—H

—Chemical shift of the CH_3 Group

0.0

Chemical shift of the -OH group

d) NOTE:

  I. Hydrogens in different chemical / electronic environments have different chemical shifts.

  II. The numerical value of the chemical shift is characteristic and gives a clue as to the proton that originates the signal.

  III. Delta is independent of the operating frequency.

$\delta = 7.1$ ppm

$\delta = 4.5 - 4.6$ ppm

D. Shielding / Deshielding

Upfield / Downfield absorption of protons is due to differences in their
   i)   Electronic environment

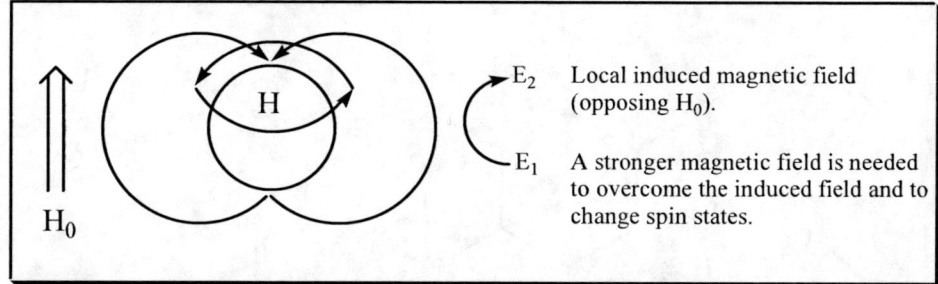

- E_2 Local induced magnetic field (opposing H_0).
- E_1 A stronger magnetic field is needed to overcome the induced field and to change spin states.

(1) Shielding of Protons

SHIELDING CAUSED BY σ ELECTRONS

(a) Circulation of sigma electrons around the nucleus creates a secondary magnetic field, which **opposes** the applied field. Higher energy is therefore necessary to bring that nucleus into resonance.

(b) Sometimes caused by the direction of the 2° magnetic field caused in compound having p electrons e.g. acetylene.

(2) Deshielding of Protons

(a) Attachment of electronegative atoms O, N, Cl, Br, F, etc..., to a methyl group causes deshielding.

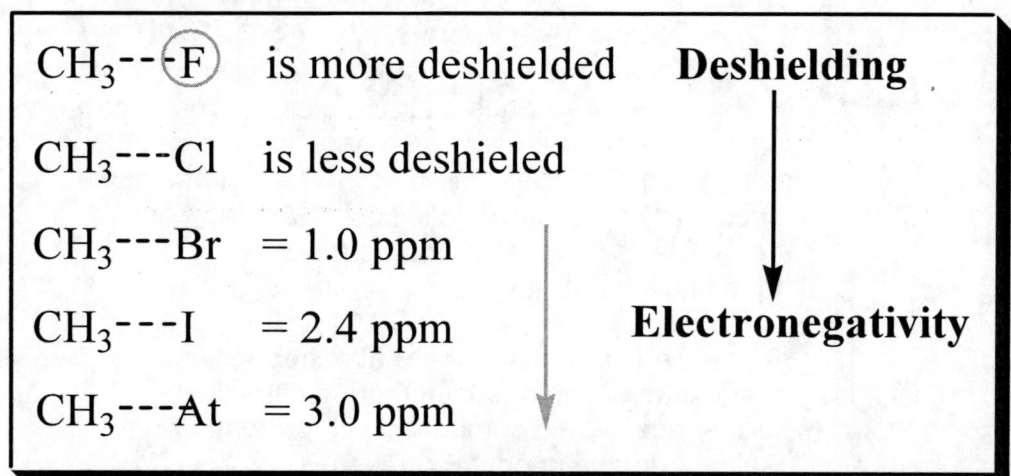

CH_3--F   is more deshielded   **Deshielding**

CH_3---Cl is less deshieled

CH_3---Br = 1.0 ppm

CH_3---I   = 2.4 ppm   **Electronegativity**

CH_3---At = 3.0 ppm

(b) Deshielding by pi electrons, **anisoptropic effects;**
Setting up a second magnetic field within the
molecular (anisotropy)

Ethylene	Benzene	Aldehyde	Amide
$\delta = 4.5 - 6.0$	$\delta = 7.1 - 7.5$	$\delta = 9.0$	$\delta = 9.0 - 10.0$

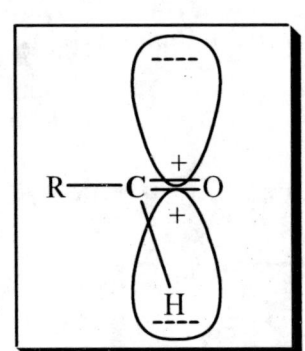

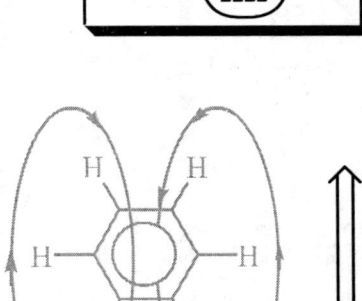

(i)   Alkene protons
   1. Anisotropy produces a magnetic field that adds to the applied field. Deshields alkene protons = 4.5 – 6.0 ppm.

(ii)   Aromatic protons
   1. Ring current equals a strong magnetic field. Aromatic protons are more shielded than alkene protons (~ 7.1 ppm).
   This is due to the fact the circulation of the six π electrons establishes a strong electrical field called "a ring current." This in turn sets up a secondary magnetic field [primary being the applied field H_0]. The spatial volume of the secondary magnetic field is such that π deshields the aromatic protons more than in alkene protons.

H_0 Thus protons that are attached to systems that can maintain or sustain a ring current are further deshielded compared with protons on isolated double bonds. Ex. Alkenes
(iii)   Aldehyde protons.
(iv)   Amide Protons

E. Chemical Equivalence

a) Chemically equivalent protons

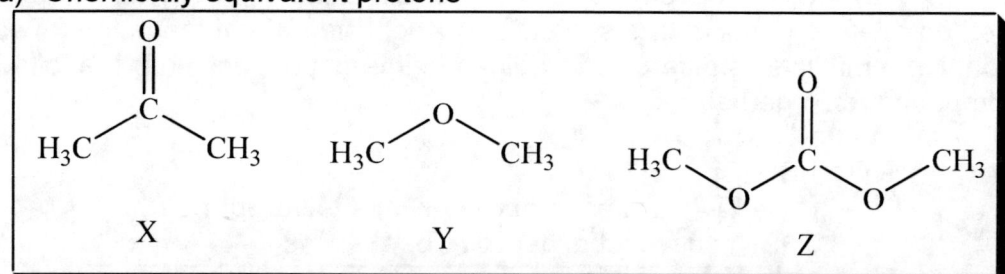

i) X, Y, and Z have one set of equivalent protons. Each compound has one signal in its H spectrum.

b) Chemically non-equivalent protons

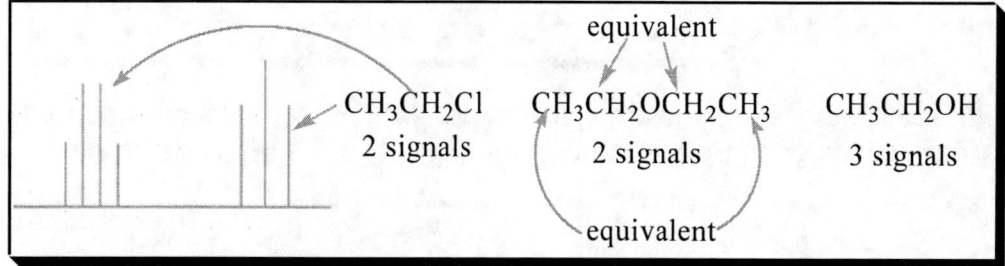

c) Test for Chemical Equivalence

To determine whether similar appearing protons are equivalent, mentally substitute another atom for each of the protons in question. If the same product is formed by imaginary replacement of either of two protons, those protons are chemically equivalent. For example, the replacement of any of the three methyl protons in ethanol by an imaginary Z atom gives the same compound; these hydrogens are chemically equivalent.

155

F. Spin – Spin Interactions

a) As mentioned before, often certain signals in a p-NMR appear as multiplets (i.e., doublets, triplets, quartets, etc...). Further, the ratio of the height of each individual signal in a multiplet is different. To spin-spin interaction, consider the following compound, chloroethane.

i) CH_3CH_2Cl
  (1) Two groups of non-equivalent hydrogens.
  (2) You would expect two peaks.

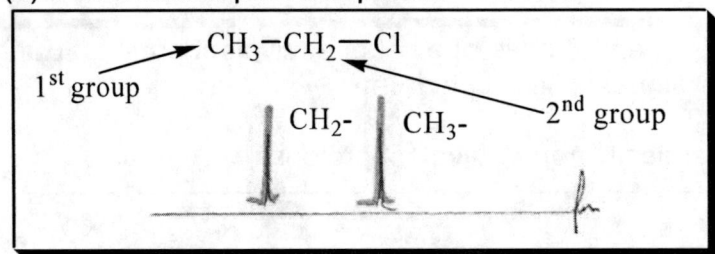

In reality, these signals are split into peaks. These peaks are due to the spin-spin interactions of non-equivalent protons.

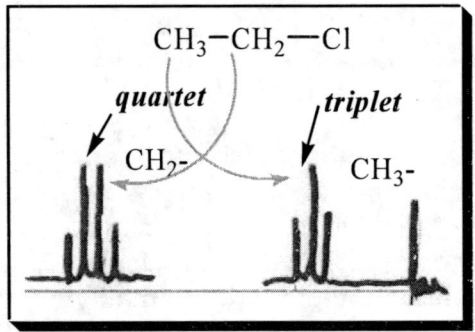

b) Spin-Spin interaction occurs between adjacent non-equivalent protons only. For example, the protons in the compound below are:

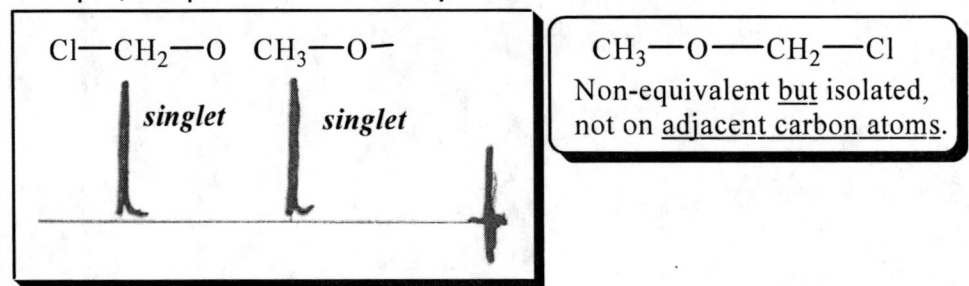

c) When the chemical shift of one nucleus is influenced by the spin of another (adjacent non-eq.) proton or a group, they are said to be coupled.

i) Consider the H_A proton. If there were no adjacent hydrogens, a singlet would occur. Since there is one adjacent hydrogen, H_B, the B_A proton is split into a doublet by the N + 1 rule. This is because the H_B proton has two spin states. One spin state is CCW, resulting in a magnetic field, which helps the applied field. This results in a slight deshielding. The other spin state is CW, resulting in a magnetic field, which is against the applied field. This results in a slight shielding of the proton.

CASE 1
DOUBLET FORMATION

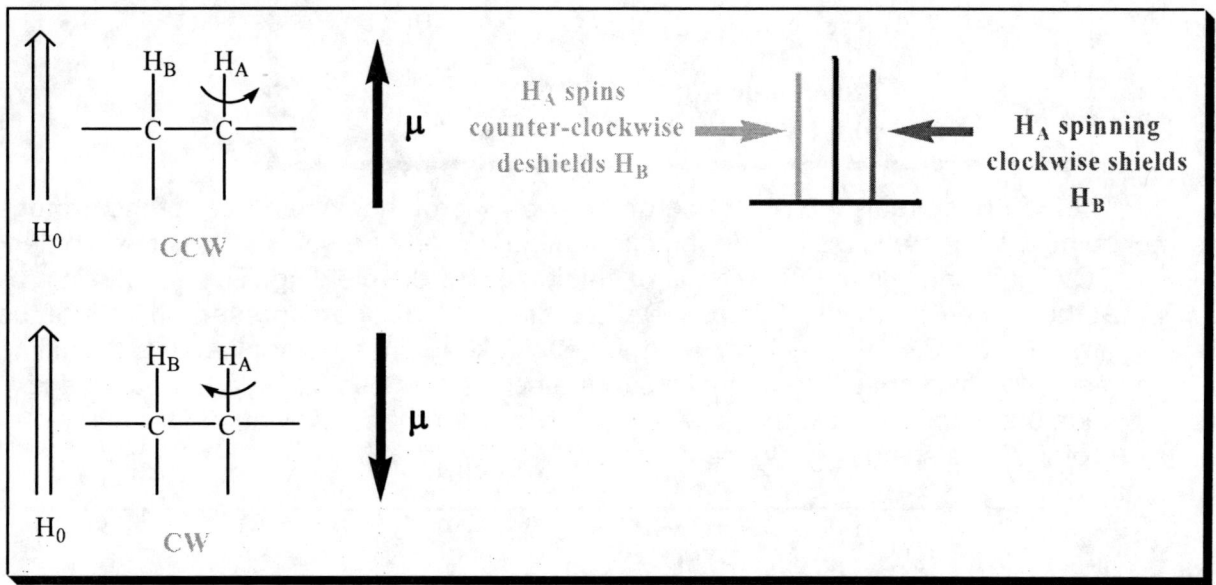

CASE 2: TRIPLET FORMATION

A triplet occurs if there are two adjacent hydrogens.

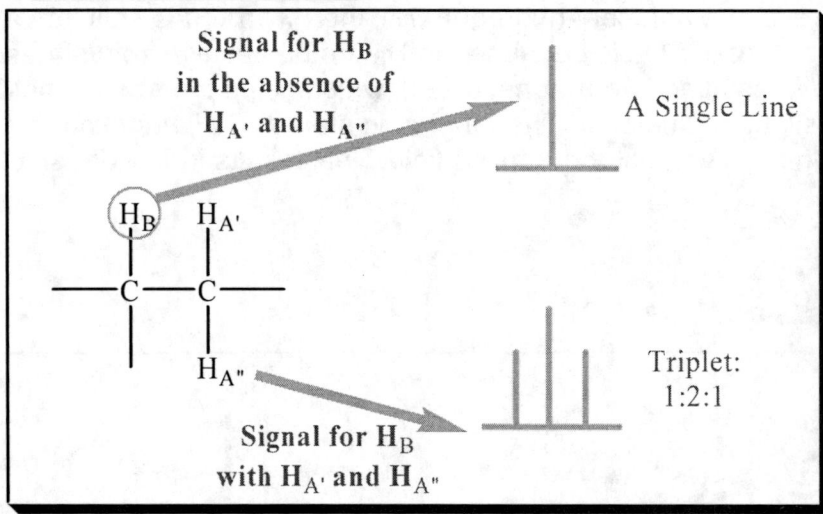

The H_B proton feels the presence of the two H_A protons. When both of the protons spin CCW, there is a slight deshielding effect. When one spins CW and the other CCW, this cancels out the effects of shielding and deshielding. This results in a peak at the same position as if there were no adjacent hydrogens present. If both of the protons spin CW, this results in a magnetic field against the applied field, resulting in a slight upfield shift. The intensities of the peaks can also be determined. In this case the ratio is 1:2:1 because there are two different configurations for the two protons, one spinning CW, the other CCW.

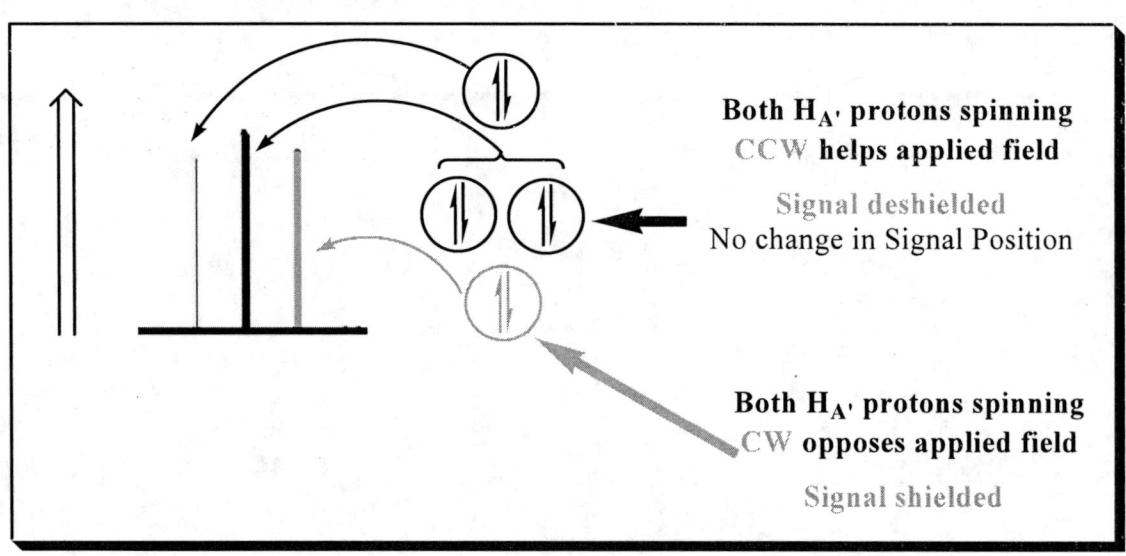

G. The Coupling Constant

Spin Spin Interaction – coupling between adjacent non-equivalent protons.

– A Summary –

1. When the chemical shift of one nucleus (proton(s)) is influenced by the spin of another nucleus (proton(s)) the two nuclei are said to be coupled.

2. Coupling occurs between non-equivalent protons on adjacent carbons only

3. Long range coupling is observed in alkenes and armoatic compounds.

4. The distance between the peaks of a multiplete (measured in Hertz) is called the coupling constant.

5. Coupling constants are represented by J.

6. The Magnetic Effect that one proton has on another depeonds on the bonds connecting the protons, but it does not depend on the strength of the external magnetic field. A spectrometer operating at 300 MHz records the same coupling constant as a 50 MHz instrument.

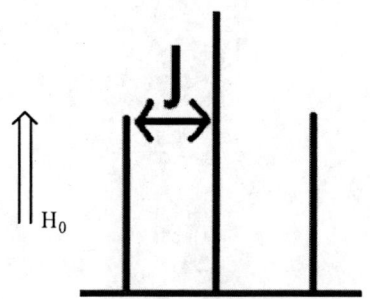

7. SOME TYPICAL COUPLING CONSTANTS.

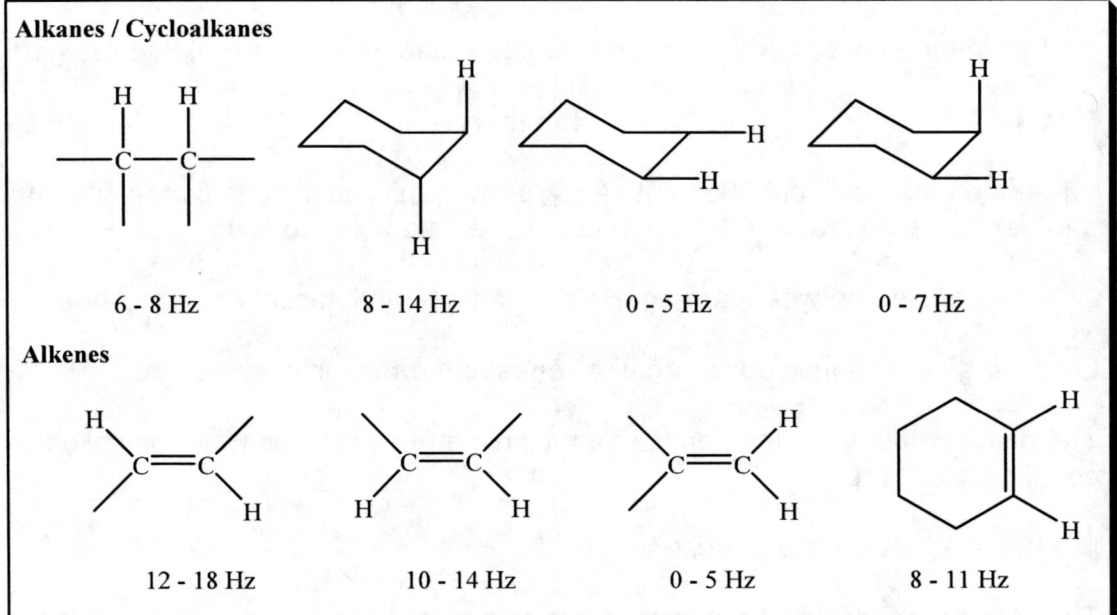

The Coupling Constant Cont'd...
Graphing analysis of spin/spin interaction
--predicting the fate of a signal due to coupling.

A) This would be the signal for H_x if it did not undergo spin-spin splitting with H_y

B) The signal is split by the two spin states of B throughout the sample.

C) The ratio of areas for the split peaks is 1: 1. Each comprises half the area of the original unsplit peak.

D) J_{xy} represents the distance between the split signal lines and represents the coupling constant, in units Hertz.

SPLIT BY TWO EQUIVALENT PROTONS

i) In the second example, H_x proton is split into a triplet by the two equivalent H_y protons.

ii) The peak for H_x is split twice, once for each H_y. The area of the center peak is twice that of the two outside peaks because it is composed of two halves of the peaks of the first splitting (one H_y), while the outermost peaks only consist of one half of each peak of the first splitting. Note that the coupling constants are all equal, since two chemically equivalent protons split H_x. H_x and H_z are non-equivalent and are both different to H_y.

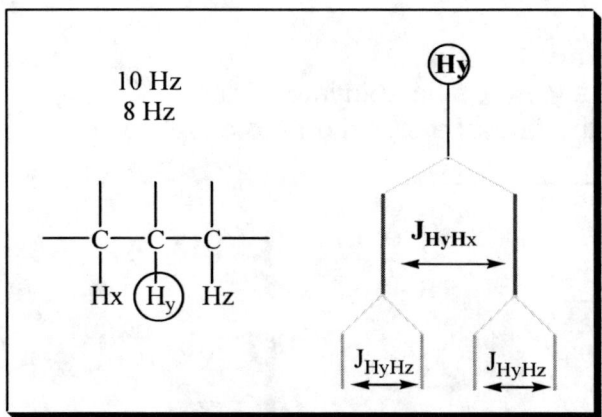

i) In the case of a hydrogen split by two non-equivalent protons, the coupling constant, J, would not be the same and the splitting pattern observed would be different. In the example above, H_y is affected by the non-equivalent H_x and H_z. Assuming in this case that J_{yx} is greater than J_{yz}.

ii) Note that the n + 1 rule does not apply this case. Instead, 4 peaks of equal area are observed since the coupling constants J_{yx} and J_{yz} are different.

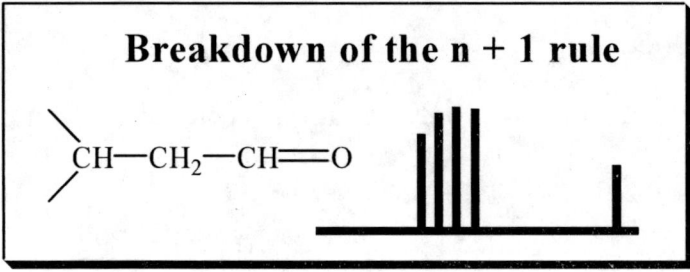

Example ③

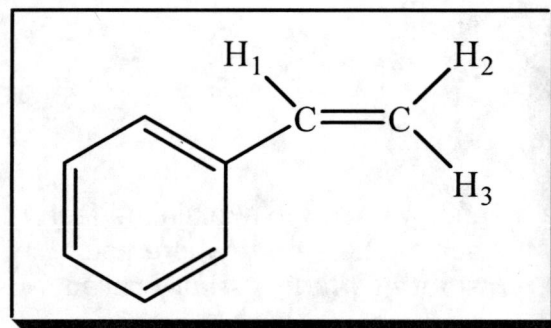

H_1 / H_3 = TRANS $J = 16$ Hz
H_1 / H_2 = CIS $J = 8$ Hz
H_2 / H_3 = GEM $J = 4$ Hz

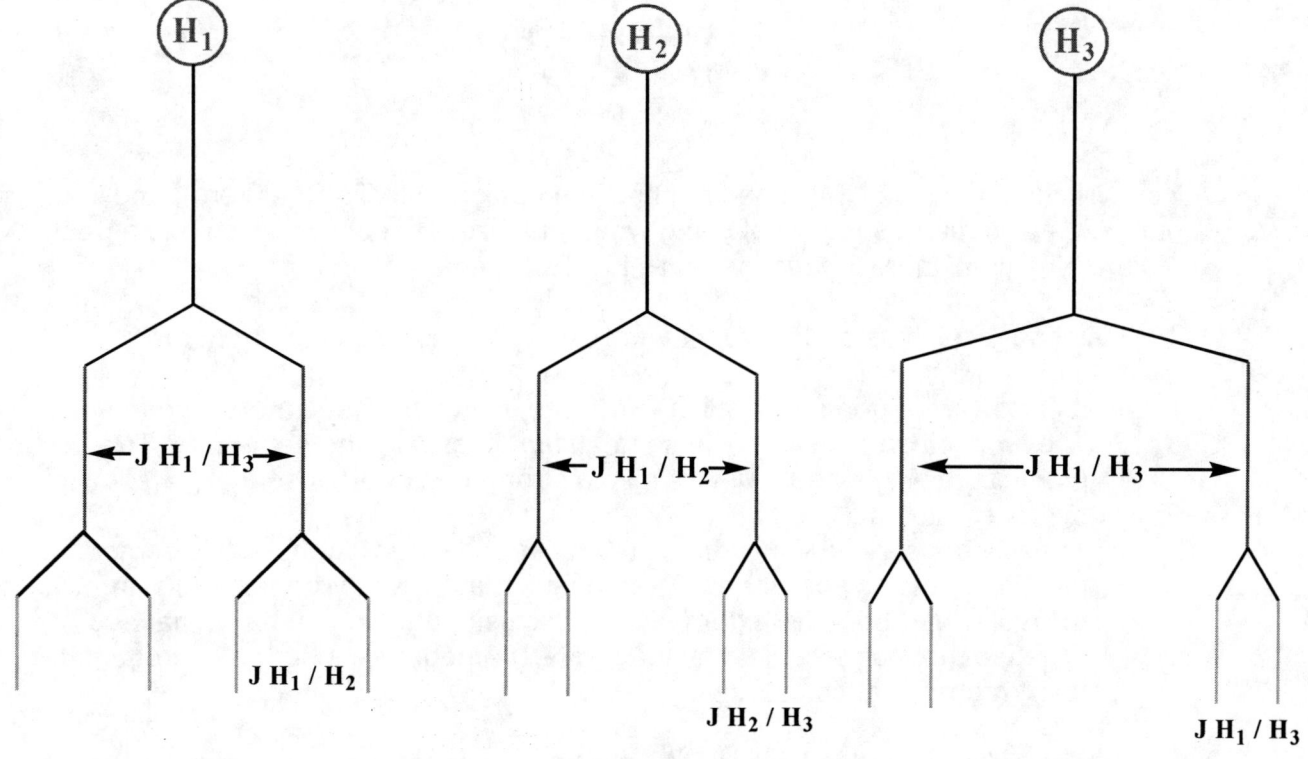

4 lines for H_1 · · · 4 lines for H_2 · · · 4 lines for H_3

A. Chemical Shifts

The need for a standard for comparison - TMS

Before we can explain what the horizontal scale means, we need to explain the fact that it has a zero point - at the right-hand end of the scale. The zero is where you would find a peak due to the hydrogen atoms in *tetramethylsilane* - usually called *TMS*. Everything else is compared with this.

$$\begin{array}{c} CH_3 \\ | \\ CH_3-Si-CH_3 \\ | \\ CH_3 \end{array}$$

You will find that some NMR spectra show the peak due to TMS (at zero), and others leave it out. Essentially, if you have to analyze a spectrum which has a peak at zero, you can ignore it because that's the TMS peak.

TMS is chosen as the standard for several reasons. The most important are:

- It has 12 hydrogen atoms all of which are in exactly the same environment. They are joined to exactly the same things in exactly the same way. That produces a single peak, but it's also a strong peak (because there are lots of hydrogen atoms).
- The electrons in the C-H bonds are closer to the hydrogens in this compound than in almost any other one. That means that these hydrogen nuclei are the most shielded from the external magnetic field, and so you would have to increase the magnetic field by the greatest amount to bring the hydrogens back into resonance.

    The net effect of this is that TMS produces a peak on the spectrum at the extreme right-hand side. Almost everything else produces peaks to the left of it.

The chemical shift

The horizontal scale is shown as δ(ppm). δ is called the *chemical shift* and is measured in *parts per million* - ppm.

A peak at a chemical shift of, say, 2.0 means that the hydrogen atoms which caused that peak need a magnetic field *two millionths less* than the field needed by TMS to produce resonance.

A peak at a chemical shift of 2.0 is said to be *downfield* of TMS. The further to the left a peak is, the more downfield it is.

Hydrogens in different environments will give different chemical shifts. Here are two common correlation charts associated with the typical ranges of chemical shifts of hydrogens of different functional groups:

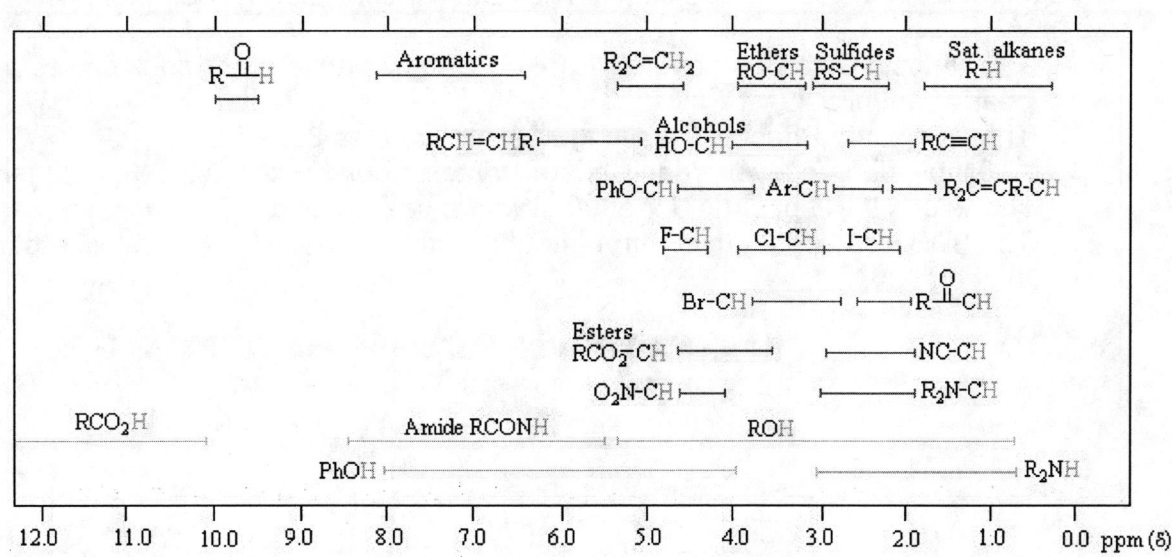

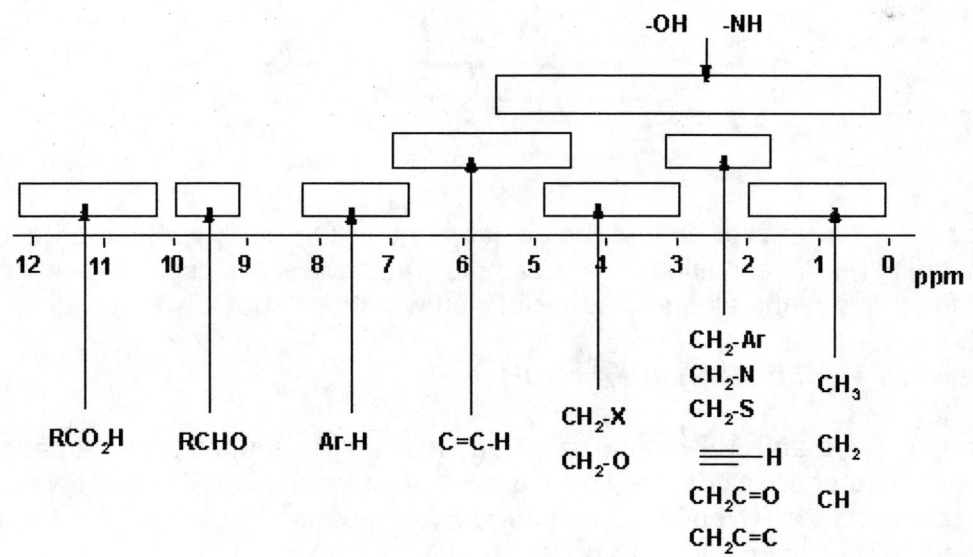

Special cases

1. Alcohols

Where is the -O-H peak?

This is very confusing! Different sources quote totally different chemical shifts for the hydrogen atom in the -OH group in alcohols - often inconsistently. For example:

- The Nuffield Data Book quotes 2.0 - 4.0, but the Nuffield text book shows a peak at about 5.4.
- The OCR Data Sheet for use in their exams quotes 3.5 - 5.5.
- A reliable degree level organic chemistry text book quotes 1.0 - 5.0, but then shows an NMR spectrum for ethanol with a peak at about 6.1.
- The SDBS database (used throughout this site) gives the -OH peak in ethanol at about 2.6.

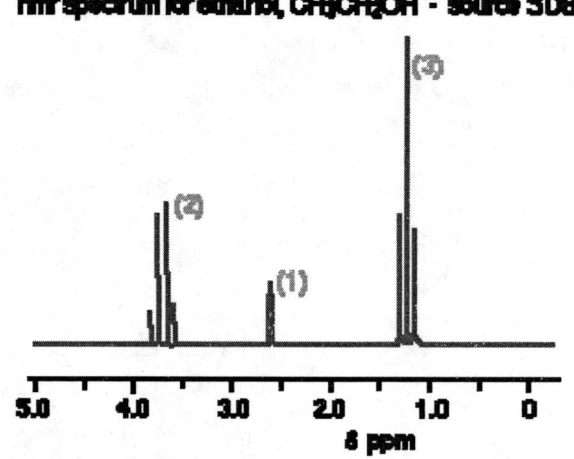

The problem seems to be that the position of the -OH peak varies dramatically depending on the conditions - for example, what solvent is used, the concentration, and the purity of the alcohol - especially on whether or not it is totally dry.

A clever way of picking out the -OH peak

If you measure an NMR spectrum for an alcohol like ethanol, and then add a few drops of deuterium oxide, D_2O, to the solution, allow it to settle and then re-measure the spectrum, the -OH peak disappears! By comparing the two spectra, you can tell immediately which peak was due to the -OH group.

The reason for the loss of the peak lies in the interaction between the deuterium oxide and the alcohol. All alcohols, such as ethanol, are very, very slightly acidic. The hydrogen on the -OH group transfers to one of the lone pairs on the oxygen of the water molecule. The fact that here we've got "heavy water" makes no difference to that.

$$CH_3CH_2\text{-}O\text{-}H + \overset{}{\underset{D}{O\text{-}D}} \longrightarrow CH_3CH_2\text{-}O^- + H\text{-}\overset{+}{\underset{D}{O}}\text{-}D$$

The negative ion formed is most likely to bump into a simple deuterium oxide molecule to regenerate the alcohol - except that now the -OH group has turned into an -OD group.

$$CH_3CH_2\text{-}O^- + D\text{-}O\text{-}D \longrightarrow CH_3CH_2\text{-}O\text{-}D + {^-}O\text{-}D$$

Deuterium atoms don't produce peaks in the same region of an NMR spectrum as ordinary hydrogen atoms, and so the peak disappears.

You might wonder what happens to the positive ion in the first equation and the OD⁻ in the second one. These get lost into the normal equilibrium which exists wherever you have water molecules - heavy or otherwise.

$$2D_2O \rightleftharpoons D_3O^+ + OD^-$$

The lack of splitting with -OH groups

Unless the alcohol is absolutely free of any water, the hydrogen on the -OH group and any hydrogens on the adjacent carbon don't interact to produce any splitting. The -OH peak is a singlet and you don't have to worry about its effect on the adjacent hydrogens.

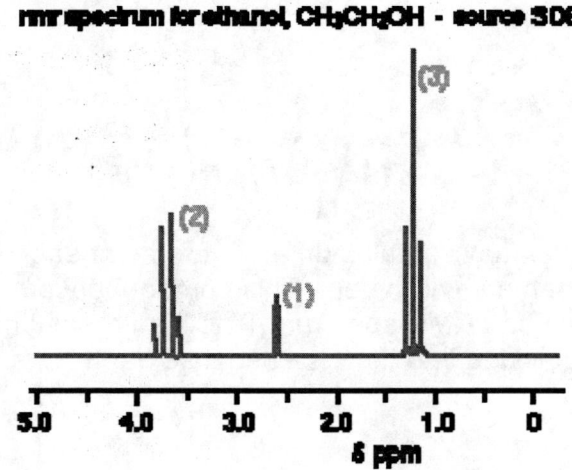

nmr spectrum for ethanol, CH_3CH_2OH - source SDBS

The left-hand cluster of peaks is due to the CH_2 group. It is a quartet because of the 3 hydrogens on the adjacent CH_3 group. You can ignore the effect of the -OH hydrogen.

Similarly, the -OH peak in the middle of the spectrum is a singlet. It hasn't turned into a triplet because of the influence of the CH₂ group.

2. Equivalent hydrogen atoms

Hydrogen atoms attached to the same carbon atom are said to be *equivalent*. Equivalent hydrogen atoms have no effect on each other - so that one hydrogen atom in a CH₂ group doesn't cause any splitting in the spectrum of the other one.

But hydrogen atoms on neighbouring carbon atoms can also be equivalent if they are in exactly the same environment. For example:

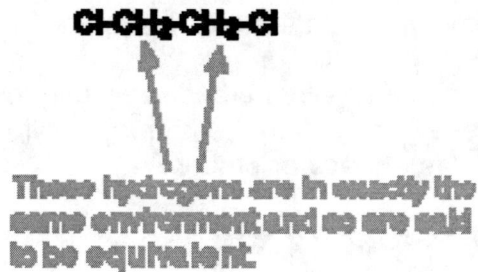

These four hydrogens are all exactly equivalent. You would get a single peak with no splitting at all.

You only have to change the molecule very slightly for this no longer to be true.

Because the molecule now contains different atoms at each end, the hydrogens are no longer all in the same environment. This compound would give two separate peaks on a low resolution NMR spectrum. The high resolution spectrum would show that both peaks subdivided into triplets - because each is next to a differently placed CH₂ group.

B. Integration

Using the areas under the peaks

The ratio of the areas under the peaks tells you the ratio of the numbers of hydrogens in the various environments. In the methyl propanoate case shown below, the areas are in the ratio of 3:2:3, which is exactly what you want for the two differently placed CH₃ groups and the CH₂ group.

A low resolution spectrum looks much simpler because it can't distinguish between the individual peaks in the various groups of peaks.

low resolution nmr spectrum for methyl propanoate, $CH_3CH_2COOCH_3$

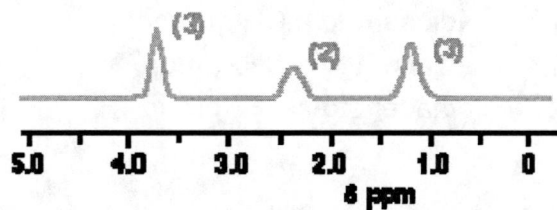

You will probably be told the relative areas under the peaks - especially if you are only looking at low resolution spectra, but it is just possible that you might have to work them out. NMR spectrometers have a device which draws another line on the spectrum called an *integrator trace* (or integration trace). You can measure the relative areas from this trace.

C. Splitting

In a high resolution spectrum, you find that many of what looked like single peaks in the low resolution spectrum are split into clusters of peaks.

For our purposes, you will only need to consider these possibilities:

1 peak	a singlet
2 peaks in the cluster	a doublet
3 peaks in the cluster	a triplet
4 peaks in the cluster	a quartet

You can get exactly the same information from a high resolution spectrum as from a low resolution one - you simply treat each *cluster of peaks* as if it were a single one in a low resolution spectrum.

But in addition, the amount of splitting of the peaks gives you important extra information.

The n+1 rule

The amount of splitting tells you about the number of hydrogens attached to the carbon atom or atoms *adjacent* to the one you are currently interested in.

The number of sub-peaks in a cluster is *one more* than the number of hydrogens attached to the adjacent carbon(s).

So - on the assumption that there is only one carbon atom with hydrogens on it next to the carbon we're interested in:

singlet   adjacent to carbon with no hydrogens attached
doublet   adjacent to a CH group
triplet   adjacent to a CH_2 group
quartet   adjacent to a CH_3 group

Using the n+1 rule
What information can you get from this NMR spectrum?

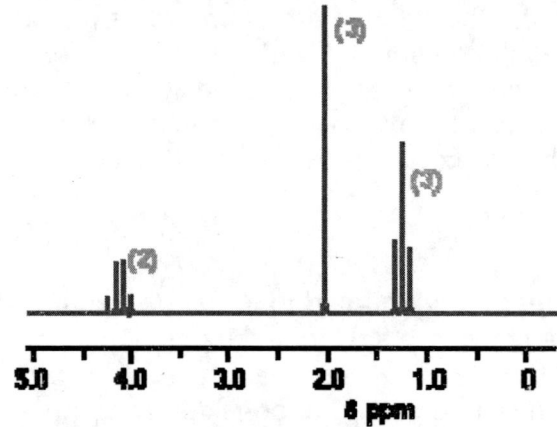

Assume that you know that the compound above has the molecular formula $C_4H_8O_2$. Treating this as a low resolution spectrum to start with, there are three clusters of peaks and so three different environments for the hydrogens. The hydrogens in those three environments are in the ratio 2:3:3. Since there are 8 hydrogens altogether, this represents a CH_2 group and two CH_3 groups.

What about the splitting?
The CH_2 group at about 4.1 ppm is a quartet. That tells you that it is next to a carbon with three hydrogens attached - a CH_3 group.
The CH_3 group at about 1.3 ppm is a triplet. That must be next to a CH_2 group. This combination of these two clusters of peaks - one a quartet and the other a triplet - is typical of an ethyl group, CH_3CH_2. It is very common. Get to recognise it!
Finally, the CH_3 group at about 2.0 ppm is a singlet. That means that the adjacent carbon doesn't have any hydrogens attached.
So what is this compound? You would also use chemical shift data to help to identify the environment each group was in, and eventually you would come up with:

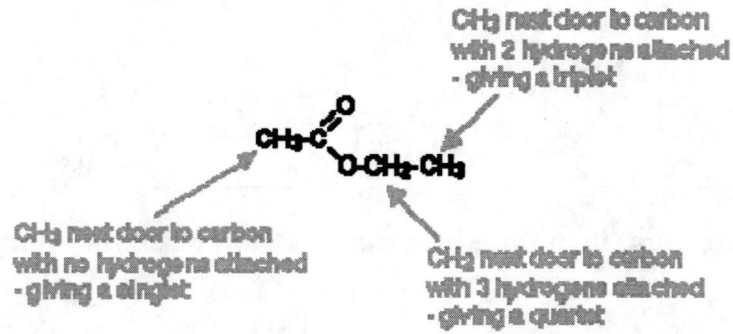

An example spectrum along with the interpretation is on the next page. More example spectra are available in Appendix C.

Problem Solving

In learning about NMR spectra, you have seen that chemical shift values can be assigned to specific types of protons, that the areas under peaks are proportional to the numbers of protons, and that nearby protons cause spin-spin splitting. By analyzing the structure of a molecule with these principles in mind you can predict the features of an NMR spectrum. Learning to draw spectra will help you to recognize the features of actual spectra. The process is not difficult if a systematic approach is used. A stepwise method is illustrated here, by drawing the NMR spectrum of the compound shown below.

1. **Determine how many types of protons are present, together with their proportions.**
   In the example above, there are four types of protons, labeled a, b, c, and d. The area ratios should be 6:1:2:3.

2. **Estimate the chemical shifts of the protons.**
   Proton b is on carbon atom bonded to oxygen; it should absorb around $\delta 3$ to $\delta 4$. Protons a are less deshielded by the oxygen, probably around $\delta 1$ to $\delta 2$. Protons c are on a carbon bonded to a carbonyl group; they should absorb around $\delta 2.1$ to $\delta 2.5$. Protons d, one carbon removed from a carbonyl, will be deshielded less than protons c, and also less than protons a, which are next to a more strongly deshielded carbon atom. Protons d should absorb around $\delta 1.0$.

3. **Determine the splitting patterns.**
   Protons a and b split each other into a doublet and a septet, respectively (a typical isopropyl group pattern). Protons c and d split each other into a quartet and a triplet, respectively (a typical ethyl group pattern).

4. **Summarize each absorption in order, from the lowest field to the highest.**

	Proton b	Proton c	Protons a	Protons d
area	1	2	6	3
chemical shift	3 – 4	2.1 – 2.5	1 – 2	1
splitting	septet	quartet	doublet	triplet

Draw the spectrum, using the information from your summary.

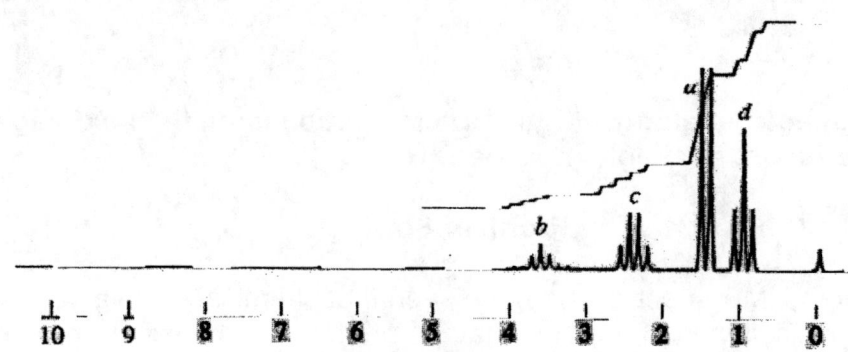

3. ^{13}C-NMR Spectroscopy

A. **Significance of ^{13}C-NMR Spectroscopy**

B. **Fundamental Difficulties Associated with ^{13}C-NMR Spectroscopy**

C. **Instrumentation**

D. **Assignment Aides**

    1. H broad band decoupling (proton decoupled)

    2. Off resonance decoupling

E. **Points to note**

    1. Non-equivalent carbons

    2. Symmetrical carbons

        ----examples

    3. Chemicals Shifts

        ----electronegativity

        ---anisotropic effects

F. **Summary**

A. Significance of ¹³C-NMR

Proton NMR spectra --- its usefulness is limited for two reasons
1. The center of interest in organic chemistry especially in natural products is not so much the protons but the carbon skeleton. The physical properties and reactivity of organic compounds are explained by the bonding states of carbon atoms.
2. Numerous organic molecules contain too few magnetically non-equivalent protons. H-NMR spectroscopy provides little or no information about the C-skeleton of such compounds.
3. Carbon-13 is the most abundant isotope of carbon; however, it does not have a spin (I=0).

$${}^{13}_{6}C \quad I = {}^{1}/_{2}$$

Carbon-12-even
Carbon--6-even
Carbon-13-odd [does have a resultant spin of 1/2 turn and therefore Carbon-13 is used as a probe for NMR Spectroscopy of organic compounds.]

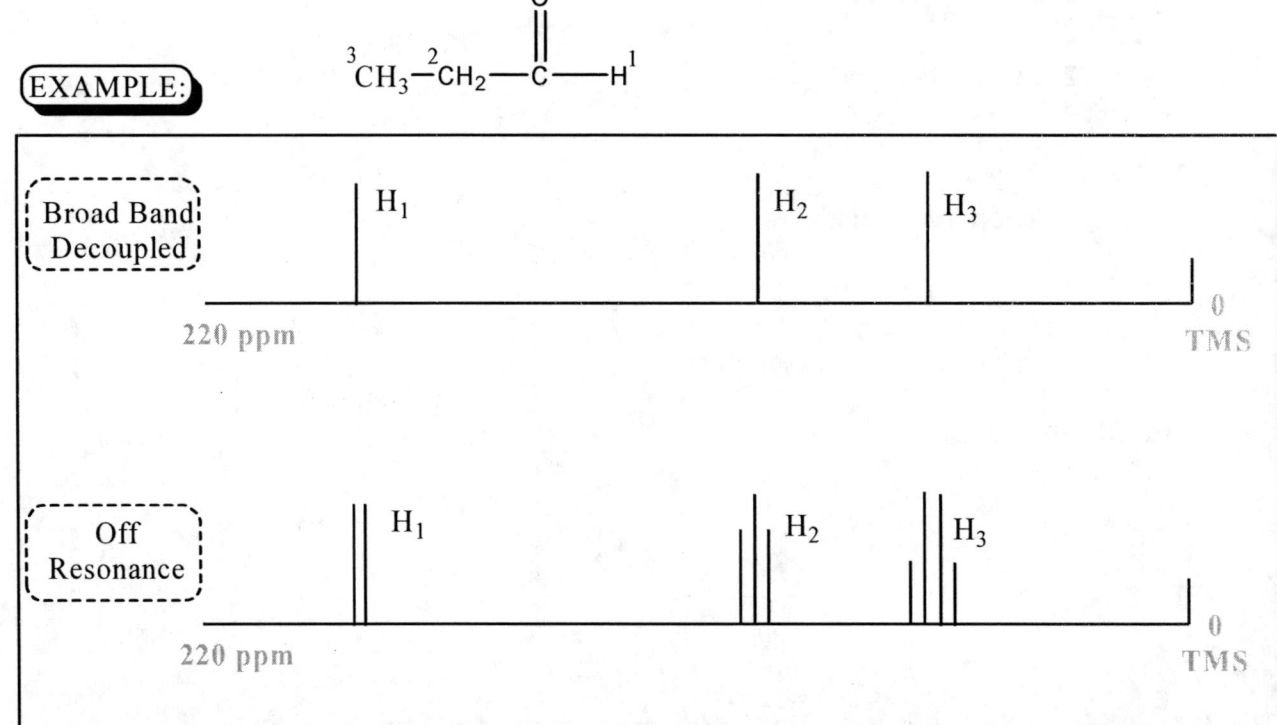

EXAMPLE: $CH_3-CH_2-C(=O)-H$ (³CH₃-²CH₂-C-¹H)

B. Fundamental Difficulties
1. Abundance of Carbon-13 in natural carbon is only 1.1%, hence large quantities of sample are necessary for observation of Carbon-13.
2. The magnetic moment of Carbon-13 is low. Consequently the resonances of Carbon-13 atoms are about 6000 times weaker than those of hydrogen.
A technique called PFT NMR spectroscopy is used (Pulse Fourier Transform)

C. Instrumentation
1 A pulse interferogram is a complex wave function resulting from the superposition of all the Lamour frequencies stimulated by the pulse.
2. Like any complex wave, the pulse interferogram can be converted under certain mathematical conditions into a spectrum of Lamour frequencies that is into the NMR spectrum itself, by Fourier transformation.
The advantage of PFT over conventional NMR is a saving of time or improvement of the signal/noise ratio by more than one order of magnitude. The sensitivity is increased if the pulse interferograms are accumulated by means of averaging computers before the Fourier transformation.

D. Assignment Aids
1. ^1H broad band decoupling {proton decoupling}.
Since geminal ^{13}C-H coupling constants [direct coupling] are very large [100-200Hz] the interpretation of ^{13}C NMR is difficult due to overlapping ^{13}C--^1H multiplets. To simplify this, broad band decoupling is used.

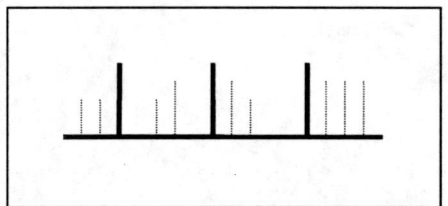 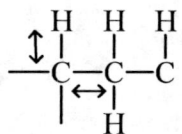 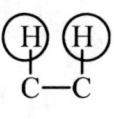

The ^{13}C sample is irradiated with a second radiofrequency H-2 in addition to the (continuous or) pulse irradiation with the field H-1.

The power of H-2 is such that irradiation covers the entire range of Lamour frequencies of the protons. All the ^{13}C-H multiplets collapse. A singlet appears for each of the non-equivalent carbon atoms.

The second RF -extra energy- causes rapid interconversions between the parallel and antiparallel spin states of the protons. As a result the ^{13}C nucleus sees only an average of the two spin states of the protons; not split, it is decoupled.

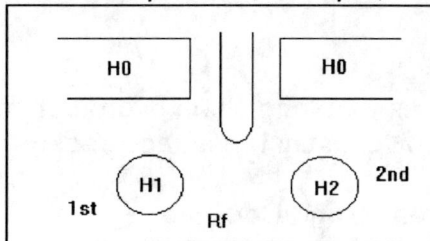

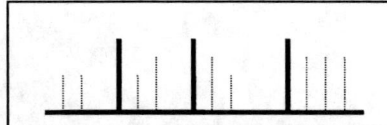 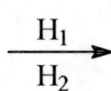 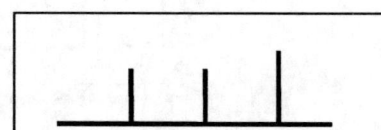

Limitation of broad band decoupling: Doesn't show us the protons on the C's. (Doesn't show the multiplicity.)

2. Off Resonance Decoupling:

H-1 broad band decoupling simplifies ^{13}C NMR spectrum and increases sensitivity; however, it suppresses valuable clues as to the assignment of the signals on the basis of the ^{13}C- 1H multiplets.

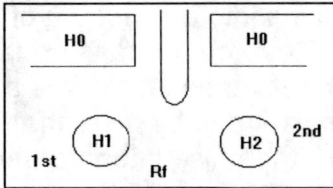

In off resonance [double resonance] the 2nd frequency is used a few hundred Hz outside the range of the frequencies of the protons instead of within the range. Vicinal and long range couplings collapse but the geminal C--H couplings remain.

N+1 rule applies: follows strictly
A spectrum of intense multiplets of first order is observed...

-quartets for methyl

-triplets for methylene

-doublets for methine

-singlets for quaternary

E. Points to note:
   1. Adjacent carbons of a long chain hydrocarbon e.g. R-[CH$_2$]$_n$--CH$_2$--CH$_2$---R' have their own distinct resonace peaks.

Broad Band Coupling

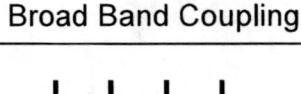

2. Carbons in the same molecule may resonate at the same chemical shift if they are equivalent by symmetry.

e.g. -1

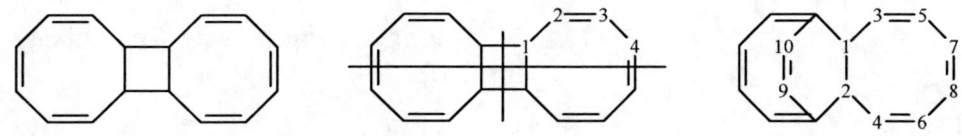

The two ortho carbons are represented by a single peak. The meta carbons give a single peak in a noise decoupled (broad band decoupled). 2≡6; 3≡5

e.g. -2 Cyclooctatetraene

Proton decoupled ^{13}C spectrum of this dimer shows four sharp lines. Fits structure # 1 which has two planes of symmetry.

1≡2, 3≡6, 4≡5

1,2 dichlorobenzene has a plane of symmetry - three types of carbons. 1,3 dichlorobenzene has a plane of symmetry with four types of carbon.

3. Electronegativity, hybridization, and anisotropic effects all affect ^{13}C chemical shifts just as they do for protons but in a more complex fashion.

4. Chemical shift positions of carbon atoms are reported by the number of ppm units they are shifted downfield from TMS. ^{13}C chemical shifts cover a wide range 0-200 ppm [0-20 ppm in ^{1}H pNMR].

F. Summary

CH₃--CH₂--CH=O

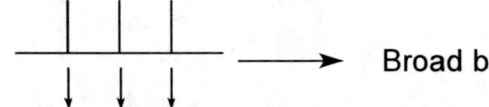 Broad band decoupling.

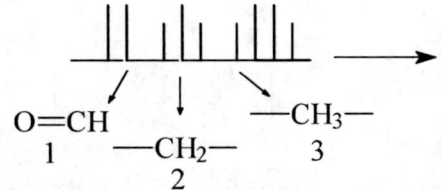 Off resonance decoupling (multiplicity n+1 rule applied).

O=CH —CH₂— —CH₃—
 1   2   3

1) The H splits the C signal into 2
2) 2 H's split C signal into 3
3) 3 H's split signal into 4

Relative positions of ^{13}C NMR absorption.

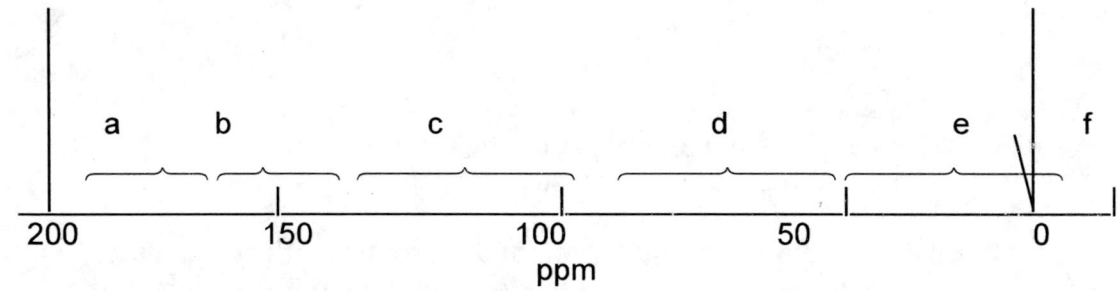

a) Aldehyde and ketone C   b) Ester, amide and carboxyl C

c) Alkene and aromatic C   d) Alkynyl C

e) Alkyl C   f) TMS

Problem Solving

Following are structural formulas for three constitutional isomers of molecular formula C₇H₁₆O and three sets of ¹³C-NMR spectral data. Assign each constitutional isomer its correct spectral data.

a) CH₃CH₂CH₂CH₂CH₂CH₂CH₂OH   1

b) CH₃C(OH)(CH₃)CH₂CH₂CH₂CH₃   2

c) CH₃CH₂C(OH)(CH₂CH₃)CH₂CH₃   3

Spectrum 1	Spectrum 2	Spectrum 3
74.66	70.97	62.93
30.54	43.74	32.79
7.73	29.21	31.86
	26.60	29.14
	23.27	25.75
	14.09	22.63
		14.08

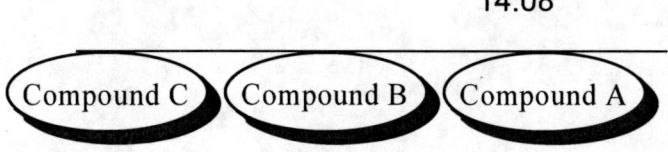

Compound C   Compound B   Compound A

CORRELATION CHART

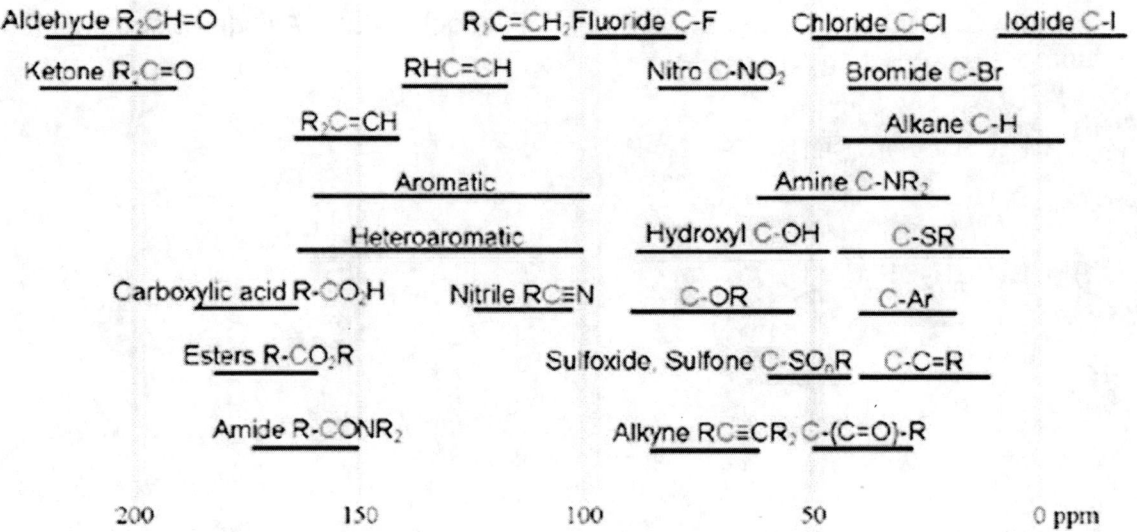

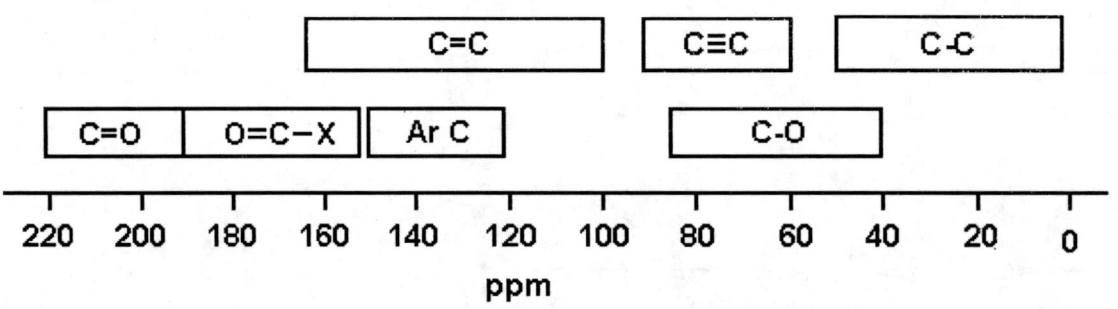

Experiment One

Identification of an Unknown Aldehyde or Ketone

OBJECTIVES

1. Prepare a 2,4-dinitrophenylhydrazone (2,4-DNP) derivative of an unknown aldehyde or ketone, recrystallize it, calculate the percent recovery, record the melting point and identify the unknown substance.

2. Prepare a semicarbazone derivative of an unknown aldehyde or ketone, recrystallize it, calculate the percent recovery, and determine the melting point and identify the unknown substance.

THEORY

1. **Characteristics of carbonyl compounds**
   a) **General description**
   - The carbonyl carbon is sp^2 hybridized with bond angles of 120°.
   - Carbon and all atoms bound directly to it lie in the same plane.

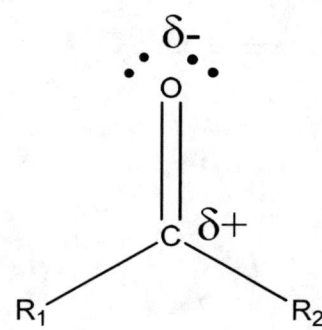

b) Bond polarization
- Due to the electronegativity of oxygen, the carbon results in a polarized carbon-oxygen bond.
- Therefore the carbon is electron-deficient known as the electrophile and is subject to nucleophilic attack.

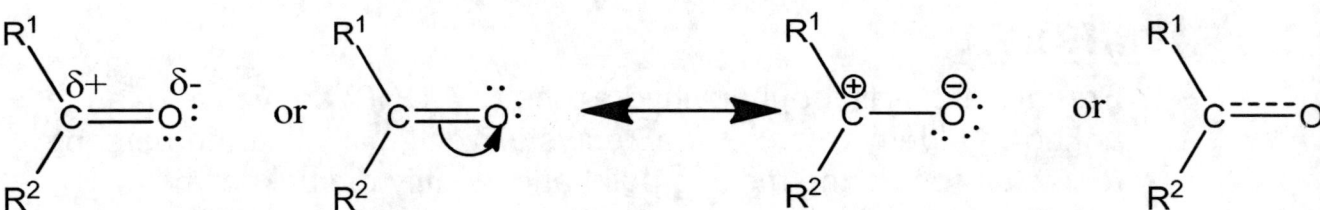

- Nucleophilic attack on the sp² hybridized carbon occurs by displaying π-bond

2. Reactions of carbonyl compounds.
- Most reactions are acid catalyzed.
- The acid will protonate the electronegative oxygen to form a carbocation.
- A carbocation is susceptible to nucleophilic addition.

General Mechanism

Addition of Nitrogen Nucleophiles to Aldehydes and Ketones

All reactions are:
1. Reversible
2. Acid Catalyzed

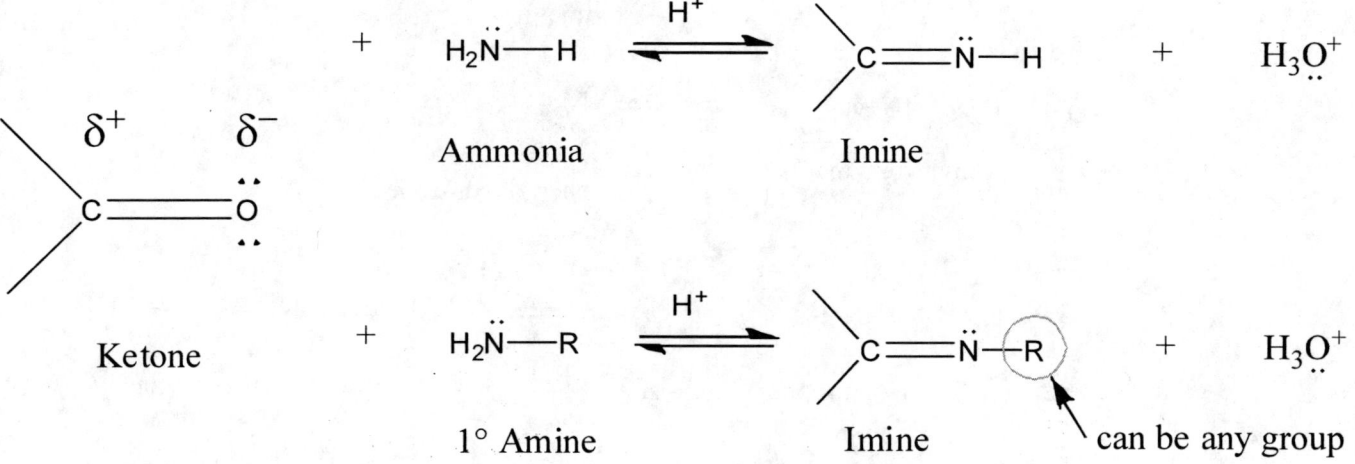

Examples of Carbonyl Reactions with Nitrogen Nucleophiles
1. Aldehydes and ketones in question for this experiment are all liquid at room temp.
    a. Cannot be distinguished via boiling point
2. Derivatization provides a solid for analysis

CHEM 332L Experiment 1: Identification of an Unknown Aldehyde or Ketone

EXAMPLES OF STABLE AMINES

1. $\text{C=O} + \text{H}_2\text{N-NH}_2 \overset{H^+}{\rightleftharpoons} \text{C=N-NH}_2 + \text{H}_3\text{O}^+$
   hydrazine → hydrazone

2. $\text{C=O} + \text{H}_2\text{N-OH} \overset{H^+}{\rightleftharpoons} \text{C=N-OH} + \text{H}_3\text{O}^+$
   hydroxylamine → oxime

3. $\text{C=O} + \text{H}_2\text{N-NH-C}_6\text{H}_5 \overset{H^+}{\rightleftharpoons} \text{C=N-NH-C}_6\text{H}_5 + \text{H}_3\text{O}^+$
   phenylhydrazine → phenylhydrazone

4. $\text{C=O} + \text{H}_2\text{N-NH-C}_6\text{H}_3(\text{NO}_2)_2 \overset{H^+}{\rightleftharpoons} \text{CH=N-NH-C}_6\text{H}_3(\text{NO}_2)_2 + \text{H}_3\text{O}^+$
   2,4-dinitrophenylhydrazine → 2,4-dinitrophenylhydrazone 2,4-DNP
   (Orange/Red Solids)

5. $\text{C=O} + \text{H}_2\text{N-NH-C(=O)-NH}_2 \overset{H^+}{\rightleftharpoons} \text{C=N-NH-C(=O)-NH}_2 + \text{H}_3\text{O}^+$
   semicarbazide → semicarbazone
   (White Solids)

The product in **Example 4** (2,4-dinitrophenylhydrazones) and the product in **Example 5** (semicarbazones) are called "derivatives" of aldehydes / ketones. They are stable, crystalline solids with a sharp melting point. These derivatives are used in analytical studies to identify unknown liquid aldehydes / ketones whose boiling point differences are insufficient to distinguish them.

CHEM 332L Experiment 1: Identification of an Unknown Aldehyde or Ketone

Derivatization
1) 2,4-dinitrophenylhydrazine (orange syrupy liquid)
    - Dissolved in $\underline{EtOH/H_3PO_4}$ will form a solid 2,4-dinitrophenylhydrazone when reacted with an aldehyde or ketone
    - The characteristic melting points may range widely in some cases.
        - When there is no definitive result, 2,4-DNP derivative data can be confirmed with semicarbazone derivative data to provide a positive ID

CHEM 332L Experiment 1: Identification of an Unknown Aldehyde or Ketone

Application

2,4-Dinitrophenylhydrazones / Semicarbazones are used to identify unknown aldehydes / ketones in practical organic chemistry.

1. Unknown aldehyde / ketone gave a boiling point of 179°C – 181°C.

2. Possible compounds within the ± boiling point range 174°C – 186°C.

KNOWN ALDEHYDE	BOILING POINT	2,4-DNP DERIV	SEMICARBAZONE DERIV
A	176 °C	214 °C	166 °C
B	183 °C	102 °C	241 °C
C	178 °C – 179 °C	186 °C	132 °C
X	174 °C – 181 °C	185 °C – 186 °C	131 °C – 132 °C

3. Prepare derivatives [solids] and record melting point.

4. The melting point of the 2,4-DNP derivative of X match closely with the melting point of the 2,4-DNP derivative of known aldehyde C (185 – 186°C versus 186°).

5. ∴ Unknown X = Known C

Thought Question: Explain mechanistically as to why 2,4-DNP's and Semicarbazones are **not** readily hydrolyzed by H_3O^+ / Heat?

2,4-DNP Derivatization Reaction Mechanism

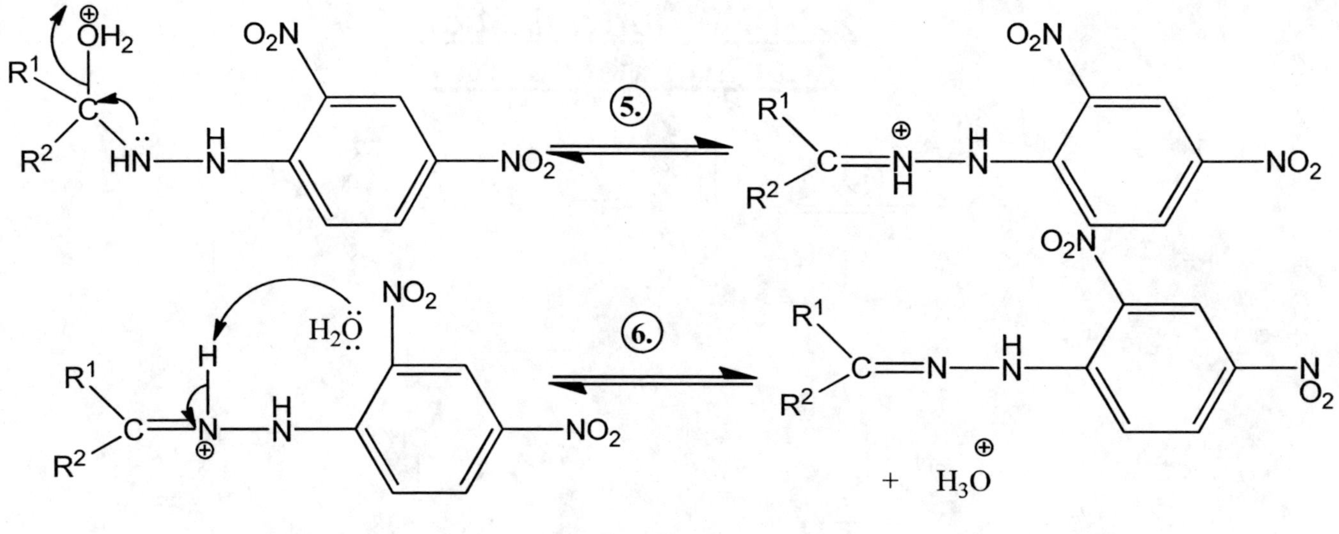

CHEM 332L Experiment 1: Identification of an Unknown Aldehyde or Ketone

Semicarbazide Derivatization

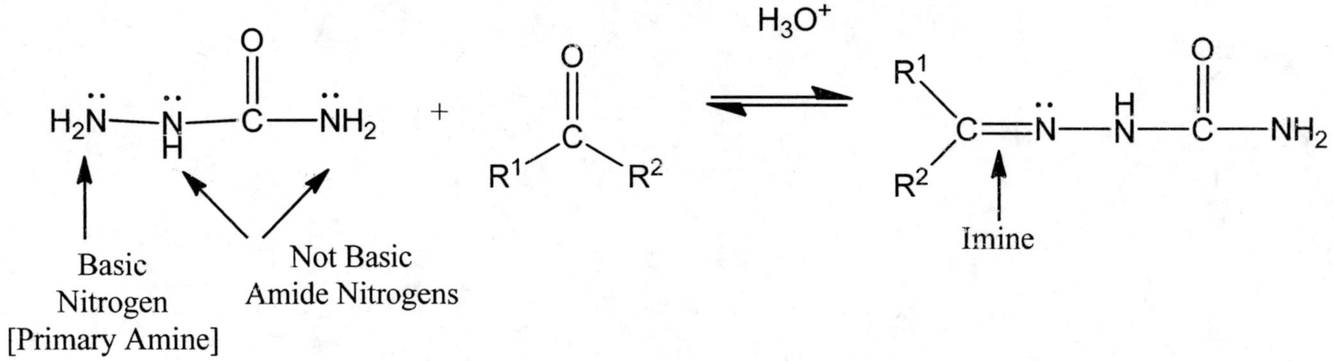

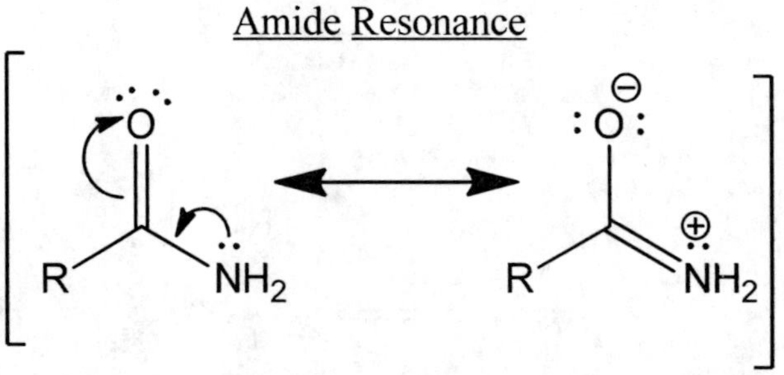

Amide Resonance

188

CHEM 332L Experiment 1: Identification of an Unknown Aldehyde or Ketone

REAGENTS

Name, Structure, MW (g/mol)	Melting °C	Boiling °C	Density g/mL	Properties	2,4 DNP Derivative(MP)	Semicarb Derivative(MP)
Benzaldehyde C_7H_6O 106.12	-26	178-179	1.045	Flammable liquid; Cherry-like scent	237 °C	216 °C
3-Heptanone $C_7H_{14}O$ 114.18	-39	147	0.818	Colorless liquid with a mild fruity odor; slightly soluble	81 °C	101 °C
Acetophenone C_8H_8O 120.15	19.6	202	1.03	Colorless liquid but solid at lower temperatures	238 °C	198 °C
Diethyl ketone $C_5H_{10}O$ 86.1334	-42	101	0.814	Flammable liquid	156 °C	138 °C
Cyclohexanone $C_6H_{10}O$ 98.1444	-47	155	0.947	Flammable Liquid Oily, white to slightly yellow liquid with a peppermint-like odor.	162°C	166°C

CHEM 332L Experiment 1: Identification of an Unknown Aldehyde or Ketone

Name, Structure, MW (g/mol)	Melting °C	Boiling °C	Density g/mL	Properties	Safety
Phosphoric acid (structure) H_3PO_4 97.99506	21 °C	158 °C	1.685	Viscous, colorless, odorless, liquid; solidifies at 70 °F hygroscopic	Corrosive to tissue and irritating to the skin, mucous membranes, upper respiratory tract and eyes; fatal if swallowed.
Ethanol (structure) C_2H_5OH 32.042	-114.1 °C	78.3 °C	0.789	Colorless liquid; pleasant alcoholic odor	Harmful by ingestion, inhalation or skin absorption; irritant of the eyes, nose throat and skin (flashback along the vapor trail may occur)
Methanol (structure) CH_3OH 32.042	-98 °C	64.6 °C	0.791	Colorless liquid with a characteristic pungent odor	May cause severe skin and eye irritation; it can be absorbed through the skin; it may cause narcosis.
Pyridine (structure) C_5H_5N 79.1012	-41.6 °C	115.2 °C	0.9819	Colorless or yellow liquid with a penetrating sickening stench	Toxic by ingestion, inhalation and contact of the skin, eyes and respiratory tract; NH_3 may cause sterility in males.
2,4-DNP (structure) $C_6H_6N_4O_4$ 198.1378	198 °C	n/a	-	Flammable liquid; orange crystals with ammonia-like odor	May cause eye irritation; FLAMMABLE; Possible mutagen
Semicarbazide hydrochloride (structure) CH_6ClN_3O 111.53309	175 – 177 °C	~185 °C	-	Very soluble in water; white crystals	Toxic; local irritant

CHEM 332L Experiment 1: Identification of an Unknown Aldehyde or Ketone

SAFETY

- ✓ Please handle all chemicals carefully, as they are potentially hazardous.
- ✓ The 2,4-DNP solution is 85% phosphoric acid. Contact with the skin is to be avoided. 2,4-DNP solution will permanently stain clothes.
- ✓ Semicarbazide solution will irritate the eyes and mucous membranes.
- ✓ There are flammable and volatile liquids in use.
- ✓ Take care to avoid broken glass and sharp objects at all times.
- ✓ Gloves will be provided when you enter the lab in order to prevent direct contact of the organic solvents with your skin. Wear goggles at <u>all</u> times.
- ✓ Pyridine is known to have sterilizing effects on males, so handle very carefully. It can give severe headaches on momentary exposure.

DISPOSAL

- ✓ Wash excess aqueous solutions down the drain with excess water.
- ✓ Dispose of the 2,4-DNP filtrate liquids in the organic liquid waste container.
- ✓ Dispose of the aldehyde / ketone derivatives in the solid organic waste container.
- ✓ Dispose of soiled gloves and paper towels in proper containers. Dispose of broken glass in the *broken glass container*.
- ✓ Check with your TA if you are in doubt concerning disposal procedures

CHEM 332L Experiment 1: Identification of an Unknown Aldehyde or Ketone

PROCEDURE

REMINDER: close and cap all reagent and waste containers

1. Obtain 2mL of your unknown from the hood and record the number of the unknown you receive.

   Preparation of 2,4-DNP Derivative:

   [1-4 Reaction]
   1. Acquire 4 mL of the 2,4-DNP stock solution from the fume hood and place it in a medium test tube.

   2. Add approximately 1 mL of the unknown compound to the test tube containing the 2,4 DNP.

CHEM 332L Experiment 1: Identification of an Unknown Aldehyde or Ketone

3. Prepare a heating apparatus by placing a beaker 2/3 full with water with a thermometer on the steam bath. Make sure that the thermometer is not touching the bottom of the beaker. Why? Maintain the temperature at 60° – 65°C by regulating the steam flow.

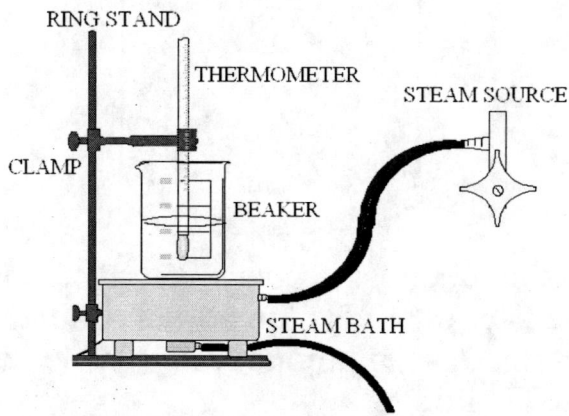

Warm water bath (Steam bath)

4. Heat the mixture for 5 - 10 minutes in the warm water bath. **[Warming promotes the forward reaction!]**

[5-7 Isolate Collect Crude Product]

5. Remove the test tube from the water bath and slowly cool the mixture to room temperature to begin crystallization (10 - 15 minutes). Then place in an ice bath until crystallization is complete. Make sure there is plenty of ice in the beaker the whole time and if needed scratching the inside of the test tube with a glass stirring rod will induce crystallization.

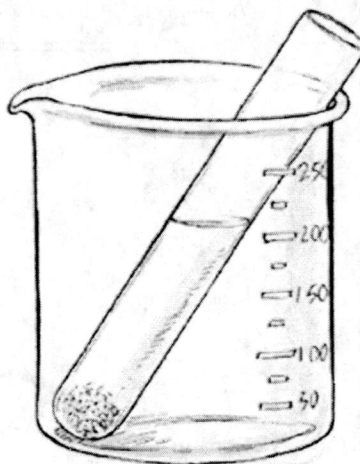

6. Isolate the solid product using suction filtration utilizing the funnel. It is important the filter paper fits the funnel exactly, otherwise product will be lost. Wash the solid product with 1mL ice cold water. Add ice cold water drop by drop.

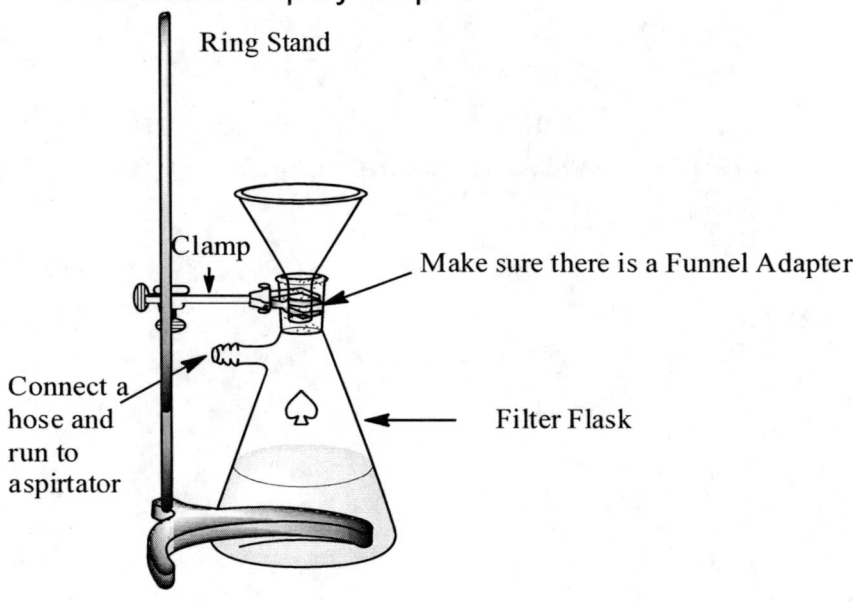

Suction Filtration Apparatus

CHEM 332L Experiment 1: Identification of an Unknown Aldehyde or Ketone

7. Weigh the crude product.

8. Transfer the product to a small test tube and recrystallize it as follows:

    a. Cover the crude 2,4-DNP derivative in the test tube with a minimal amount of warm ethanol (C_2H_5OH).
    b. Heat gently while agitating on the steam bath.
    c. <u>Slowly</u> add ethanol (C_2H_5OH). Dropwise until all the solid has dissolved.
    Note: some soluble and/or insoluble impurities may not dissolve regardless; proceed to step 9.

9. Cool the hot solution in the test tube slowly to room temperature, then place in an ice bath, to complete the crystallization.

10. Isolate the solid via suction filtration. Wash the product with 1 mL of ice cold ethanol (C_2H_5OH). [Pure Product]

11. Collect the pure product, dry, calculate the percent recovery, and determine the melting point.

CHEM 332L Experiment 1: Identification of an Unknown Aldehyde or Ketone

Preparation of Semicarbazone Derivative:

[1-5 Reaction]

1. Place 1 mL of the stock semicarbazide hydrochloride solution in a medium test tube.

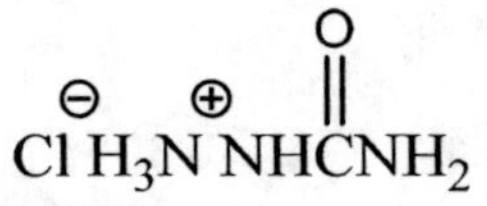

$$Cl^{\ominus} \; H_3N^{\oplus} \; NHCNH_2 \; (\text{with } C=O)$$

2. Add approximately 1 mL of the unknown compound and about 2 mL of methanol.

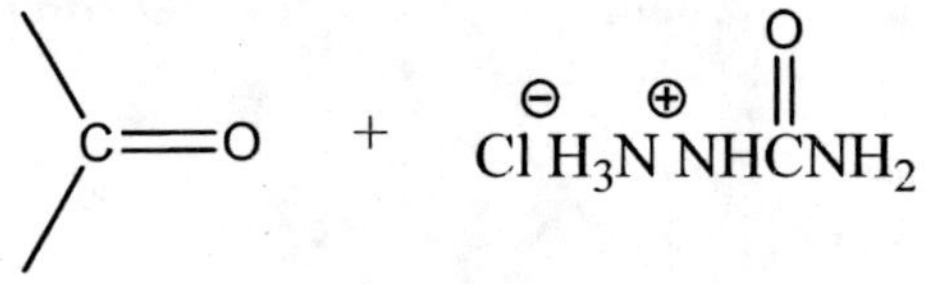

$$\text{>C=O} \;+\; Cl^{\ominus} \; H_3N^{\oplus} \; NHCNH_2(=O)$$

3. Add 20 drops of pyridine an organic base. *This step must be performed IN THE HOOD!* Warm the solution gently on the steam bath for approximately 8-10 minutes.

A. Pyridine + $Cl^{\ominus} H_3N^{\oplus} NHC(=O)NH_2$ (Salt) $\longrightarrow$ $H_2N\text{-}NHC(=O)\text{-}NH_2$ (Free Base) + Pyridinium$^+$ $Cl^{\ominus}$

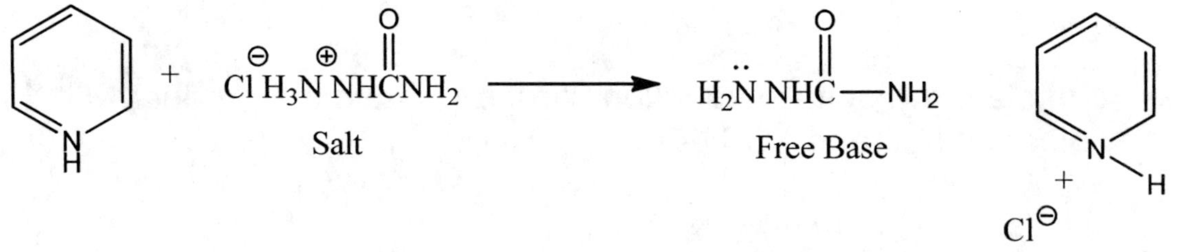

B. >C=O + $H_2N\text{-}NHC(=O)\text{-}NH_2$ $\rightleftharpoons$ $\text{>C=N-NH-C(=O)-NH}_2$

Semicarbazone

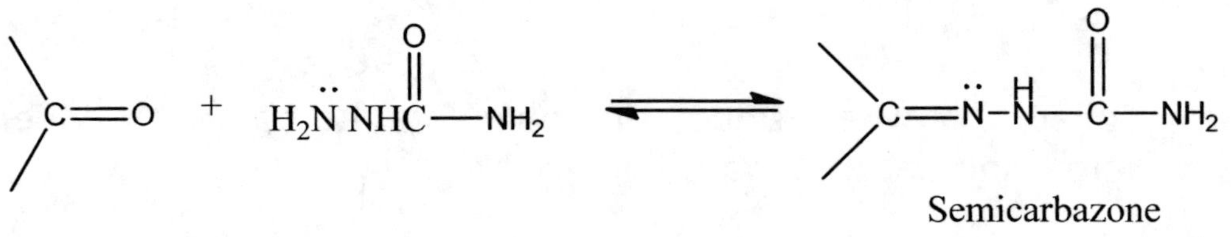

CHEM 332L Experiment 1: Identification of an Unknown Aldehyde or Ketone

4. Cool the solution in the test tube slowly to room temperature. Place the tube in an ice bath to complete the crystallization. Scratch the inside of the tube with a glass rod to induce crystallization if needed.

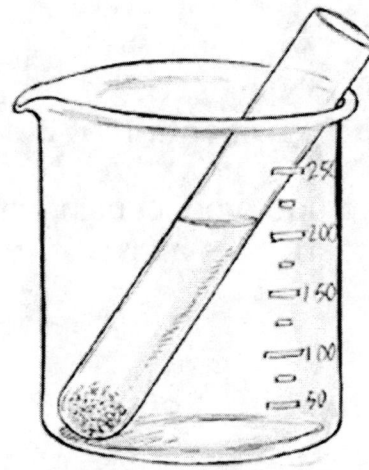

5. Isolate the cool mixture and collect the crude solid product by suction filtration.

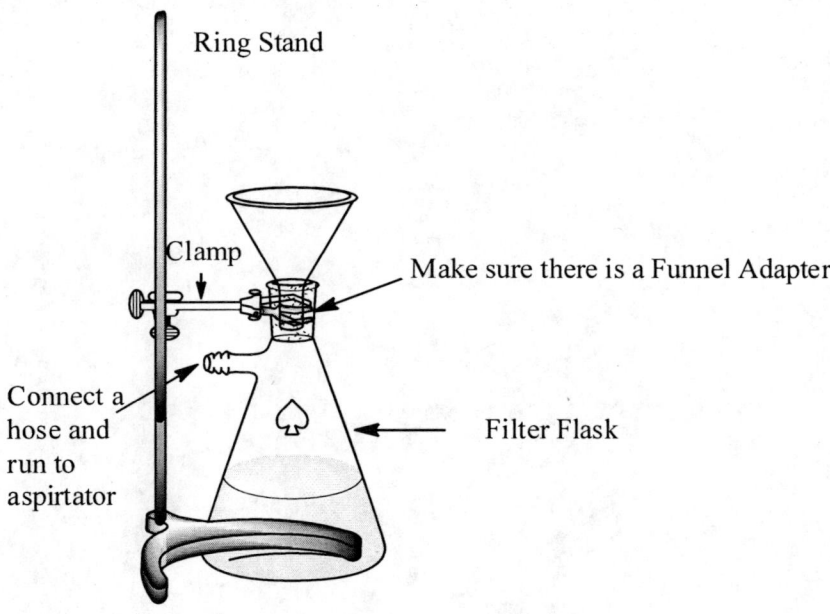

CHEM 332L Experiment 1: Identification of an Unknown Aldehyde or Ketone

[Step 6 Isolation of the Crude Product]

6. Wash the crude product with 10 drops of <u>ice cold water</u> followed by the dropwise addition of 5 drops of <u>ice cold methanol</u>. Why does the methanol have to be ice cold?

[Step 7 Purification of the Product]

7. Recrystallize the crude product using the instructions from steps 8-9 of the 2,4 DNP derivative above.

[Step 8 Identifying Pure Solid by Recording the Melting Point]

8. Collect the pure product by suction filtration, air dry, calculate the percent recovery, and determine the melting point.

OBSERVATIONS

Record observations that should be included in your report.

Own Observations:

CHEM 332L Experiment 1: Identification of an Unknown Aldehyde or Ketone

Results

UNKNOWN: = _____

Mass Crude 2,4 DNP product = _____ g

Mass Pure 2,4 DNP product = _____ g

Percent recovery of 2,4 DNP derivative = _____ %

Melting point of 2,4 DNP derivative = _____ °C

Percent difference = _____ %

Percent recovery of semicarbazone derivative = _____ %

Mass Crude semicarbazone product = _____ g

Mass Pure semicarbazone product = _____ g

Melting point of semicarbazone derivative = _____ °C

Percent difference = _____ %

Identity of Unknown = _____

Calculations

CONCLUSIONS

Experiment Two

A Grignard Reaction

OBJECTIVES
1. Prepare the Grignard reagent from bromobenzene and magnesium shavings.
2. Synthesize triphenylmethanol from benzophenone and phenylmagnesium bromide.
3. Calculate theoretical yield.
4. Record the mass and determine the percent yield.
5. Determine melting point.

THEORY
The Grignard reaction is a carbon-carbon bond-forming reaction by which almost any alcohol can be synthesized from a Grignard reagent and a carbonyl compounds.

The Grignard reagent is an organometallic compound – a compound that contains a polar carbon-metal bond.

1. **Formation of the Grignard reagent**
   The Grignard reagent is formed by reacting an alkyl halide, in particular a bromide, with magnesium metal in anhydrous diethyl ether, which is the aprotic solvent (a liquid medium for the reaction).

   a. General reaction:

$$\underset{\text{alkyl halide}}{\text{R—Br}} + \underset{\text{magnesium metal}}{\text{Mg}} \xrightarrow{\text{Et}_2\text{O}} \underset{\substack{\text{alkyl magnesium bromide} \\ \text{(organomagnesium compound)}}}{\text{R—Mg—Br}} \longleftrightarrow \overset{\ominus}{:}\text{R} + \overset{\oplus}{\text{Mg}}\text{—Br} + \text{heat}$$

General free-radical mechanism:

Step 1:   R—Br ⟶ Ṙ + Ḃr

Step 2:   :Ṁg + Ḃr ⟶ Ṁg—Br

Step 3:   Ṙ + Ṁg—Br ⟶ [R—Mg—Br ⟷ :R⁻ ⁺Mg—Br]

b. Reactivity and properties

The Grignard reagent is a strong nucleophile and a strong base. Since it is a strong base, it will react with all protons that are more acidic than those found on alkenes and alkanes. Thus, Grignard reagents react easily with water, alcohols, thiols, etc. to regenerate the alkane. The reaction conditions must remain anhydrous for the reaction to work. The solvents are usually ethers which are aprotic. What is the difference between a protic solvent and an aprotic solvent?

:R⁻ ⁺Mg—Br + H—OH ⟶ R—H + HO—Mg—Br

c. Reaction of bromobenzene with magnesium to form Grignard reagent

Bromobenzene + Magnesium →(diethyl ether) Phenylmagnesium bromide ↔ (phenyl carbanion) + Mg^+—Br + heat

The function of diethyl ether is twofold.
    i) The aprotic solvent
    ii) Diethyl ether is used to stabilize the Grignard reagent.

Phenylmagnesium bromide dietherate

2. Reactions of Grignard reagents to give Alcohols

a. General reaction (phenylmagnesium bromide with benzophenone)

Phenylmagnesium bromide + Benzophenone →(1) ether, 2) HCl/H_2O) Triphenylmethanol + $Mg^{2+}Br^-Cl^-$ magnesium salts

b. Reaction mechanism

Phenylmagnesium bromide + **Benzophenone** → (ether) → **magnesium alkoxide** (the magnesium salt of an alcohol) → **Triphenylmethanol** + **magnesium salts** (Mg^{2+}Br$^-$Cl$^-$)

The magnesium alkoxide is a salt that is insoluble in ether. In a simple acid-base reaction the magnesium alkoxide is reacted with acidified ice water to give the covalent, ether-soluble alcohol and the ionic water-soluble magnesium salt (Mg^{2+}Br$^-$Cl$^-$).

Possible Side Reaction

The reaction of phenylmagnesium bromide with unreacted bromobenzene produces biphenyl, which is the primary impurity in this lab. It is very important that the bromobenzene be added to the reaction mixture slowly so it will react with the magnesium and not be present in high concentration to react with previously formed Grignard reagent. Biphenyl is removed very easily because it is more soluble in hydrocarbon solvents than triphenylmethanol. What would be a good solvent to separate the biphenyl from the product? Think – is biphenyl a polar compound or non-polar compound?

Bromobenzene + **Phenylmagnesium bromide** → **Biphenyl** + MgBr$_2$

Note: Magnesium metal has a coating of oxide on the outside. Cautiously crushing the metal in diethyl ether can expose a fresh surface area necessary for initiating the reaction.

CHEM 332L Experiment 2: A Grignard Reaction

REAGENTS

Name, Structure, MW (g/mol)	Melting (°C)	Boiling (°C)	Density (g/mL)	Properties	Safety
Magnesium metal Mg 24.31	648	1090	1.738	Reactive metallic solid	FLAMMABLE; highly flammable, high burning material
Diethyl ether $C_4H_{10}O$ 74.12	-116	34.6	.706	Colorless liquid with a characteristic, sweet, ether odor	Moderately toxic to humans by ingestion; absorbs through skin; FLAMMABLE; **Keep away from Heat Sources of Ignition!**
Bromobenzene C_6H_5Br 157.01	-31	156	1.495	Mobile, combustible liquid, with an aromatic odor.	Irritant of the skin, eyes and mucous embranes and can be absorbed through the skin; upon decomposition it emits toxic fumes HBr gas; high vapor concentrations may be anesthetic and or somnolent
Benzophenone $C_{13}H_{10}O$ 182.22	49	305	1.11	White crystals	IMMEDIATELY flood affected skin with water while removing and isolating all contaminated clothing. Gently wash all affected skin areas thoroughly with soap and water; if eye contact, remove contact lens and rinse with water
Hydrochloric acid HCl 36.46	-30	100	0.909	Corrosive material	Highly corrosive irritant of the skin, eyes and mucous membranes; CORROSIVE
Magnesium sulfate $MgSO_4$ 120.3626	1124	-	2.66	HYGROSCOPIC; Drying agent; White powder	Dessicant
Triphenylmethanol $C_{19}H_{16}O$ 260.33	160-163	360	-	Crystalline, white powder	Irritating to the eyes, skin and respiratory tract

CHEM 332L Experiment 2: A Grignard Reaction

SAFETY

- ✓ Both chemicals are flammable, irritants, and toxic, so make sure that the apparatus is air tight. **Diethyl ether is extremely flammable. Handle with extreme care near heating devices!!! It has a flash point of ~ 40° C.**
- ✓ **DO NOT** add boiling chips to a hot liquid. A superheated liquid may boil over or even explode.
- ✓ As a hot heating mantle is in use, extra precaution must be taken.
- ✓ The apparatus can also get quite hot during use.
- ✓ Acids such as HCl are in use and are potentially dangerous.
- ✓ Take care to avoid broken glass and sharp objects at all times.
- ✓ Gloves will be provided when you enter the lab in order to prevent direct contact of the organic solvents with your skin. Wear goggles at <u>all</u> times.

DISPOSAL

- ✓ Dispose of product (triphenylmethanol) in the solid organic waste container. DO NOT POUR DOWN THE DRAIN WITH WATER.
- ✓ Organic liquids such as bromobenzene, benzophenone, and ether should disposed of in the liquid organic waste container.
- ✓ Dispose of soiled gloves and paper towels in the appropriate containers.
- ✓ Dispose of broken glass in the *broken glass container*.

CHEM 332L Experiment 2: A Grignard Reaction

PROCEDURE

REMINDER: Close and cap all reagent and waste containers.

Formation of the Grignard Reagent
1. Weigh about 0.5 g of crushed Mg turnings and place them in a 100 mL round bottom flask (RBF).

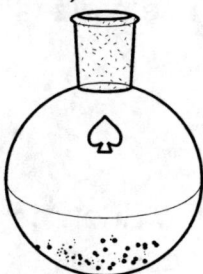

2. Prepare the Mg by adding a few mL of diethyl ether into the RBF (enough to cover the Mg turnings fully) and boil off all of the ether by steam bath while avoiding getting moisture into the system.

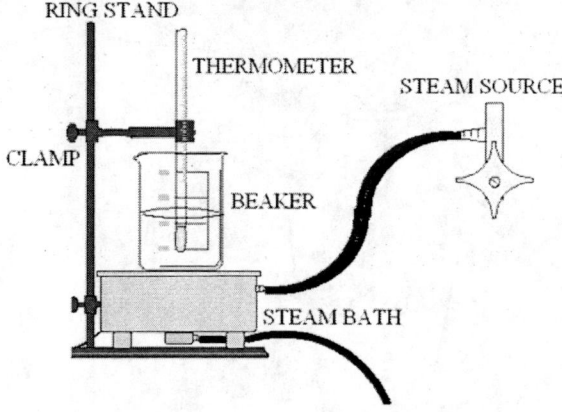

3. Pre-weigh a 10 mL graduate cylinder and measure about 2.1 mL of bromobenzene (MW=157.0), then reweigh the cylinder to find the mass of bromobenzene.

4. Transfer the bromobenzene into a 125 mL Erlenmeyer flask and use aluminum foil to cover the flask to prevent moisture from entering.

CHEM 332L Experiment 2: A Grignard Reaction

5. Take the used graduated cylinder and add 10 mL anhydrous ether, then transfer this into the Erlenmeyer flask containing bromobenzene.

6. Stir this mixture well, use a glass stirring rod.

7. Pour the contents of the Erlenmeyer flask into a separatory funnel. Make sure that the stopcock is closed before pouring this in.

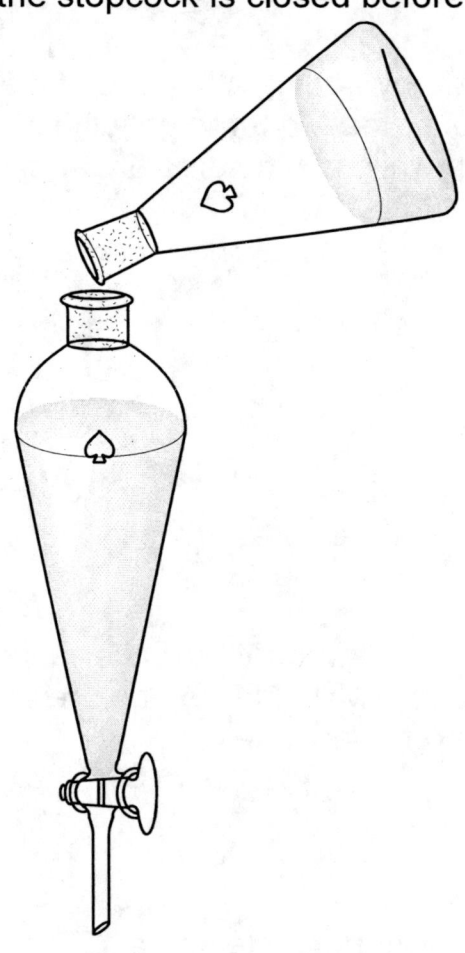

CHEM 332L Experiment 2: A Grignard Reaction

8. Release about half of the contents of the separatory funnel into the RBF (eyeball this).

9. Slowly bring the contents of the RBF to a boil (make sure the RBF never touches hot plate), bubbles will appear at the surface.

10. The contents of the RBF boil on its own and they will turn into a grayish/cloudy coloring before turning light brown (should take no longer than 25 minutes).

11. If no color change occurs, stir with a glass stirring rod to help along the reaction.

12. Once the light brown color is achieved, add the 2nd half of the separatory funnel content slowly over a period of 5 minutes so that boiling continues gently.

13. If boiling stops, add more bromobenzene. If boiling is too vigorous, lower the heat source. The Mg should be disintegrating as boiling continues.

14. Add 1.0 mL ether to the separatory funnel after the addition is complete and then release the contents into the RBF. Remove the funnel and place the drying tube in its place.

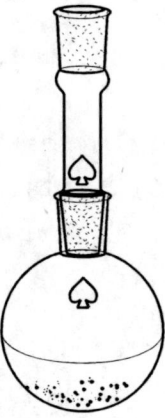

15. If the Mg has not dissolved fully, gently heat in reflux for up to another 15 minutes (a few small pieces is okay).

16. Allow the mixture to cool to room temperature.

Addition of Benzophenone
1. Weigh 2.4 g of Benzophenone and add to an Erlenmeyer flask.

2. Add 9 mL of ether into the same flask, swirl the mixture until all of the benzophenone has dissolved, and then pour mixture into a separatory funnel.

3. Add this solution into the RBF quickly, the solution should turn a rose/red color then a cloudy color.

CHEM 332L Experiment 2: A Grignard Reaction

4. Rinse the Erlenmeyer flask that contained the benzophenone solution with 5 mL ether and add this to the mixture via the separatory funnel.

5. After the addition is complete, allow the mixture to come to room temperature. As solidification occurs, and adduct should form slowly, this will also make it hard for the magnetic stirrer to stir. You will have to use a spatula to further stir the mixture.

6. Remove the RBF and place a stopper on top of it. Stir the mixture occasionally and the adduct should be fully formed after about 15 minutes.

Hydrolysis
1. Pour RBF contents into a 125 mL Erlenmeyer flask.

2. Add enough 6M HCl to the reaction mixture to neutralize it.

3. You should see the formation of 2 layers (ether on top, aqueous on the bottom). If 3 layers are present, add more ether.

4. Pour the contents of the flask into a separatory funnel and shake well, the 2 layers should form (Remember to vent frequently in between periods of shaking).

5. Drain off the aqueous layer in a beaker and then drain the ether layer into an Erlenmeyer flask.

6. Add 5-10 mL of ether into the beaker containing the aqueous layer.

Separation/Drying
1. Pour the contents of the beaker into the separatory funnel and repeat the shake and separation method again, discarding the aqueous layer and adding the upper ether layer into the reserved Erlenmeyer flask from step 5 in Hydrolysis (to recover as much product as possible).

2. Add 1.0 g of the drying agent $MgSO_4$ to remove H_2O, decant into a beaker using filter paper to keep solid product while allowing the liquid into the beaker.

3. Put the beaker on a steam bath on low heat to boil off the ether. As the ether evaporates, the remaining mixture varies from a brown oil to a colored solid mixed with an oil (biphenyl and triphenylmethanol).

4. Remove the biphenyl by slowly adding about 10 mL of petroleum ether.

5. Heat the mixture slightly, stir and allow to cool to room temperature.

6. Use suction filtration to remove the triphenylmethanol and rinse it with small portions of petroleum ether.

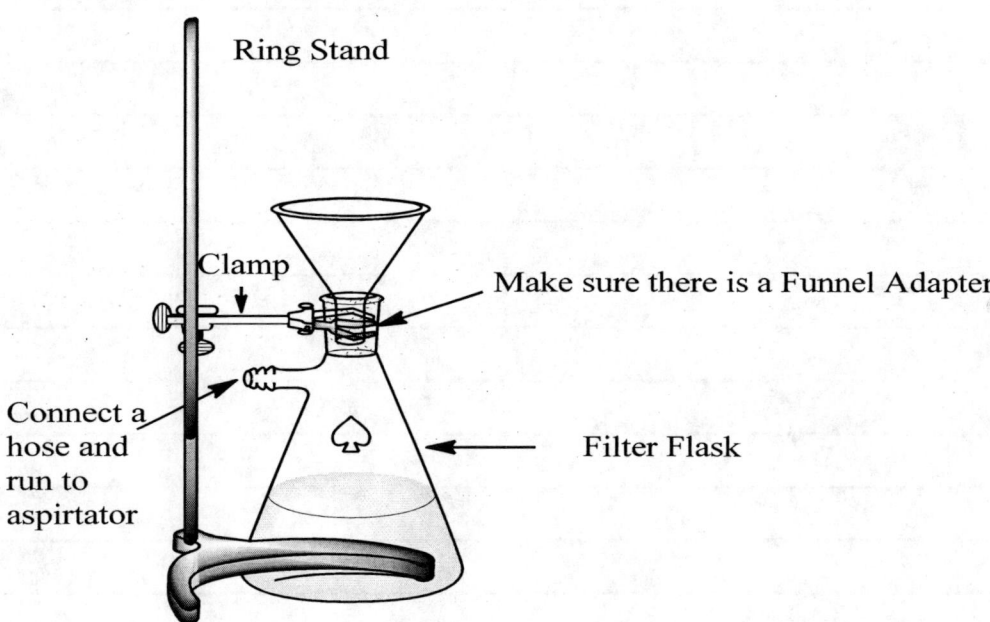

7. Air-dry the solid, weigh, calculate % yield.

$$\% \text{ Yield} = 100 \times \frac{\text{Actual Yield}}{\text{Theoretical Yield}}$$

CHEM 332L Experiment 2: A Grignard Reaction

OBSERVATIONS

1. Describe the appearance and properties of the starting materials, as well as any observations from the experiment such as color change, temperature change, and bubbling.

2. What do these changes mean?

Own Observations:

_____
_____
_____
_____
_____
_____
_____
_____
_____
_____
_____
_____
_____
_____
_____
_____
_____
_____
_____

Results

Mass of benzophenone = _____ g

Theoretical Yield of triphenylmethanol = _____ g

Mass of product (triphenylmethanol) = _____ g

Percent Yield = _____ %

Melting Point = _____ °C

Percent Difference = _____ %

Calculations

CONCLUSIONS

Experiment Three 3
An Aldol Condensation

OBJECTIVES

1. Synthesize dibenzalacetone from benzaldehyde and acetone via a mixed aldol condensation reaction.
2. Isolate the product and recrystallize from 95% ethanol.
3. Calculate the theoretical yield and percent recovery.
4. Record the mass, determine melting point, and calculate percent yield.

THEORY

1. **The Aldol Condensation Reaction**
    - Condensation Reaction: a reaction that results in the expulsion of a water molecule, or some other small, stable molecule
    - Aldol Reaction: the acid- or base-catalyzed condensation reaction of an aldehyde or ketone with another aldehyde or ketone to form a β-hydroxy ketone or an α-β unsaturated ketone
        - Very useful in carbon-carbon bond formation

2. **Base-Catalyzed Aldol**
    - Base will abstract an α-hydrogen, forming an enolate
    - Enolate (nucleophile) will react with carbonyl carbon (electrophile)

[Benzaldehyde resonance structures]

Polar Carbonyl Group
Carbon Atom is Electropile

[Benzaldehyde structure with α-carbon labeled]

The a-carbon has no removable hydrogen therefore benzaldehyde acts as the acceptor

- The reaction of an aldehyde with a ketone using sodium hydroxide is an example of a mixed aldol condensation reaction. If the aldehyde does not have any alpha hydrogens, the reaction is called a Claisen-Schmidt reaction.
- Dibenzalacetone is easily prepared by condensation of acetone with two equivalents of benzaldehyde. If only one equivalent is present, then only a monobenzalacetone will be formed.
- The aldehyde carbonyl group is more reactive than the ketone and will therefore react rapidly with the enolate anion of the ketone to give a β-hydroxyketone, which then undergoes a base-catalyzed dehydration to give benzalacetone. Ethanol is used in this experiment as the solvent to dissolve benzaldehyde, the starting material, and also benzalacetone, the intermediate.

CHEM 332L Experiment 3: Aldol Condensation Reaction

3. General Reaction
- **Claisen-Schmidt Reaction**
    - No α-hydrogens, symmetrical ketone yields only one possible product

Polar Carbonyl Group
Carbon Atom is Electropile

The a-carbon has no removable hydrogen therefore benzaldehyde acts as the acceptor

$$2 \text{ PhCHO} + CH_3COCH_3 \xrightarrow{NaOH} \text{Ph-CH=CH-CO-CH=CH-Ph}$$

Benzaldehyde + Acetone (2-Propanone) → Dibenzalacetone

CHEM 332L Experiment 3: Aldol Condensation Reaction

a. Enolate Formation: Mechanism for the formation of an enolate anion.

$CH_3C(=O)-CH(H)(H) + OH^- \rightleftharpoons [CH_3C(=O)-CH_2^- \leftrightarrow CH_3C(O^-)=CH_2] + H_2O \rightleftharpoons$

(electron withdrawing carbonyl; acidic α-carbon H; resonance-stabilized enolate anion)

$CH_3C(O^-)=CH_2 + H-O-H \rightleftharpoons CH_3C(OH)=CH_2 + OH^-$
(an enol)

Hydrogens attached to α-carbon atoms are acidic due to the polar electron withdrawing effect of the carbonyl group. A strong base removes acidic α-hydrogens forming an enolate anion which is resonance stabilized. The enolate anion is also a strong nucleophile.

The Aldol Condensation can also be done in the presence of catalytic amounts of an acid.

Acetone $\xrightarrow{H_3O^+}$ Enol $\rightleftharpoons$ β–hydroxy Ketone $\xrightarrow{-H_2O}$ α,β- Unsaturated Ketone

b. Mechanism for the base catalyzed Aldol Condensation reaction

STEP 1: $CH_3C(=O)-CH(H)(H) + OH^- \rightleftharpoons [CH_3C(=O)-CH_2^- \leftrightarrow CH_3C(O^-)=CH_2] + H_2O$

STEP 2: $CH_3C(=O)-CH_2^-$ (nucleophile) + PhCHO (electrophile) $\rightleftharpoons CH_3CCH_2CH(O^-)Ph$ (alkoxide)

STEP 3: alkoxide + H-O-H $\rightleftharpoons CH_3CCH_2C_\beta(OH)(H)Ph$ + OH⁻ (a β-hydroxy ketone)

STEP 4: $OH^- + CH_3C(=O)-C(H)(H)-C(H)(OH)Ph \rightarrow CH_3C(=O)-CH=CH-Ph$ + H_2O + OH^-

Benzalacetone

Then repeat the whole process again.

222

CHEM 332L Experiment 3: Aldol Condensation Reaction

a. Synthesis of Dyes

Dibenzalacetone is a yellow solid that has a sharp melting point. Ketones and aldehydes may be used to form other organic dyes. One such experiment is the synthesis of Indigo, a deep blue dye.

2 o-nitrobenzaldehdye + Acetone →(1) NaOH/H$_2$O) Indigo

REAGENTS

Name, Structure, MW (g/mol)	Melting °C	Boiling °C	Density g/mL	Properties	Safety
Sodium hydroxide NaOH 39.99	318	1390	2.13	Corrosive material; colorless, odorless solid; HYGROSCOPIC	Highly corrosive irritant of the skin, eyes and mucous membranes; CORROSIVE
Ethanol C_2H_5OH 46.07	-114.1	78	0.789	Colorless liquid; pleasant alcoholic odor	Harmful by ingestion, inhalation or skin absorption; irritant of the eyes, nose and throat, and skin; flashbacks along the haphazard vapor trails
Benzaldehyde C_7H_6O 106.12	-26	178-179	1.045	Flammable liquid; cherry-like scent	Can cause contact dermatitis and allergic reactions in susceptible individuals; narcotic in high concentrations, and when heated to decomposition it emits toxic fumes
Acetone C_3H_6O 58.0798	-94.3	56.2	0.7857	Colorless liquid with a fragrant, mint-like odor; hardening and dehardening tissues; flammable.	Toxic by ingestion and inhalation and irritant of the eyes, mucous membranes, nose, throat and upper respiratory tract; rapidly penetrates skin; absorbed through the lungs; narcotic in high concentrations
Dibenzalacetone $C_{17}H_{14}O$ 234.297	104-107	-	-	Yellowish crystals	-

CHEM 332L Experiment 3: Aldol Condensation Reaction

SAFETY

- Ethanol, acetone, and benzaldehyde are flammable, irritant, and toxic, so make sure that they are kept away from all heat sources
- **DO NOT** add boiling chips to a hot liquid. A superheated liquid may boil over or even explode.
- As a hot heating mantle is in use, extra precaution must be taken.
- Take care to avoid broken glass and sharp objects at all times.
- Gloves will be provided when you enter the lab in order to prevent direct contact of the organic solvents with your skin. Wear goggles at <u>all</u> times.
- As with all acids and bases, extra precaution should be taken when sodium hydroxide is in use.

DISPOSAL

- Wash excess aqueous solutions down the drain with excess water.
- Organic solvents are to be placed in pre-designated organic liquid waste container separate from other organic liquid waste.
- Dispose of soiled gloves and paper towels in trash. Dispose of broken glass in the *broken glass container*.
- Dispose of both ethanol/acetone and benzaldehyde solutions in the liquid organic waste container. DO NOT POUR DOWN THE DRAIN WITH WATER.
- Dispose of soiled gloves and paper towels in trash.
- Dispose of broken glass in the *broken glass container*.
- Consult TA if unsure.

PROCEDURE

REMINDER: Close and cap all reagent and waste containers.

1. Place 2 mL of 10% sodium hydroxide solution in a small test tube.

2. Add 1.6 mL of ethanol/acetone solution (4.4% acetone, 95.6% ethanol) to the test tube.

$$CH_3C(=O)-CH_2-H + OH^- \xrightarrow{1.6\ mL\ C_2H_5OH} \left[CH_3C(=O)-\overset{..}{C}H_2^- \rightleftharpoons H_3CC(-O^-)=CH_2 \right]$$

Enolate

CHEM 332L Experiment 3: Aldol Condensation Reaction

3. Add 0.2 mL of benzaldehyde to the test tube.

$$CH_3\overset{O}{\underset{}{C}}CH_3^{\ominus} + Ph-CHO \longrightarrow Ph-\underset{H_2C-C-CH_3}{\overset{H}{\underset{\underset{O}{\|}}{C}}-O^{\ominus}} \xrightarrow{H-OH}$$

Nucleophilic Addition of Enolate to the Aldehyde

4. Cap the test tube with a test tube stopper immediately and agitate vigorously.

(intermediate mechanism arrows, with OH⁻ abstracting proton (a), leading to)

$$Ph-CH=CH-\overset{O}{\underset{}{C}}-CH_2-(H) \xrightarrow{Repeat}$$

$$Ph-CH=CH-\overset{O}{\underset{\|}{C}}-CH=CH-Ph$$

Product

| 4.4 % Acetone 95.6% C₂H₅OH + 2mL 10% NaOH Solutions | → Add 0.2mL of Benzaldehyde | → Dissolve Slowly | → Pale Yellow Solution |

5. Continue to agitate the test tube occasionally for the next 30 minutes, to complete the reaction. If a solid does not form after 5 minutes, remove the cap and scratch the inside of the tube with a glass stirring rod.

$$H_3C-\overset{O}{\underset{\|}{C}}-\underset{a}{CH_2}-H \xrightarrow{OH^{\ominus}} H_3C-\overset{O}{\underset{\|}{C}}-\overset{..}{CH_2}^{\ominus}$$

$$H_3C-\overset{O}{\underset{\|}{C}}-\overset{..}{CH_2}^{\ominus} + Ph-CHO \longrightarrow Ph-\underset{\underset{O^{\ominus}}{}}{CH}-CH_2-\overset{O}{\underset{\|}{C}}-CH_3 \longrightarrow$$

To complete the reaction

6. After the 30 minutes, centrifuge the test tube for a few minutes to let the product sediment. Decant the supernatant liquid in the test tube containing the product.

7. Add 3 mL of water to the test tube, cap and swirl vigorously. Let it stand for 3 minutes. Centrifuge the tube and decant the supernatant liquid. Repeat this step. It is important to remove as much water as possible. Failure to do so will result in difficulty during recrystallization, as an oily substance may form.

8. Filter the product by suction filtration and air dry for a few minutes.

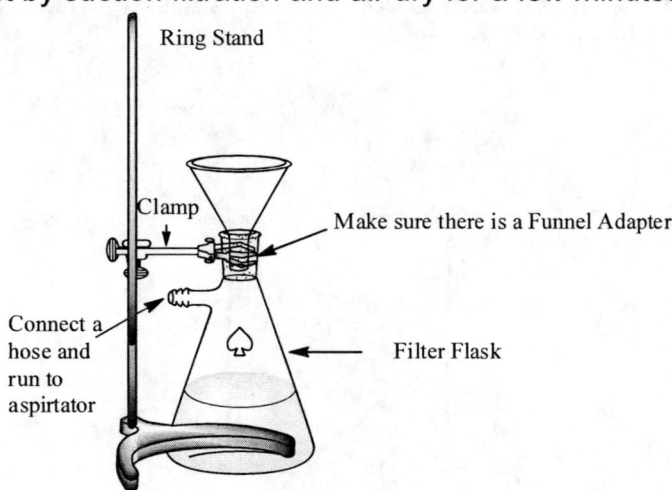

CHEM 332L Experiment 3: Aldol Condensation Reaction

9. Record the mass of the crude product.

10. Place the crystals in a test tube and add enough ethanol to barely cover the crystals. Warm the tube on a heating mantle. Insert a boiling chip (of known mass) to promote smooth boiling.

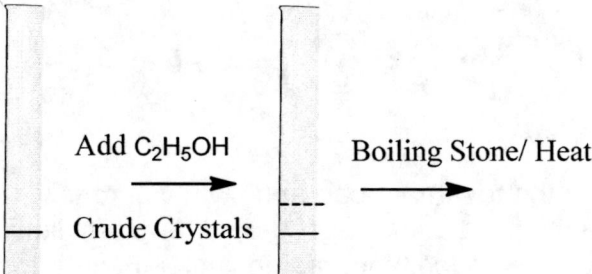

11. Heat until the solid product dissolves.

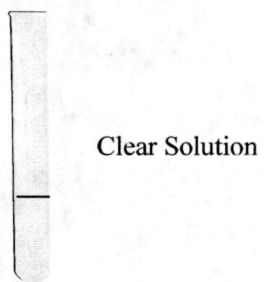

12. Remove the tube from the heating mantle and allow the solution to cool to room temperature.

13. Place the test tube in an ice bath to complete crystallization. Make sure to keep plenty of ice in the beaker!

CHEM 332L Experiment 3: Aldol Condensation Reaction

14. Filter and collect the pure crystals using suction filtration. What liquid should you wash the product with?

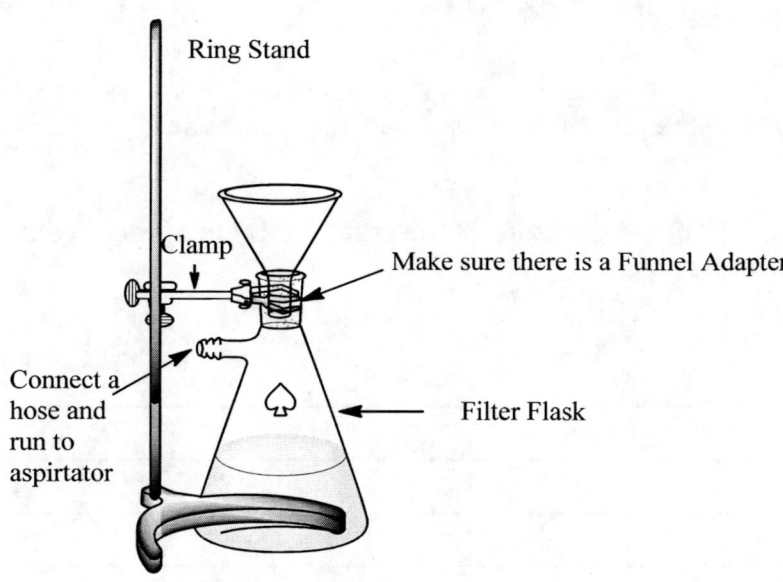

15. Air dry the product by spreading the solid evenly on a dry piece of filter paper. Cover the pure crystals on the filter paper with an inverted beaker. Leave the product for 10 minutes to air dry.

16. Record the mass and calculate the percent recovery. Determine the melting point and calculate the percent yield.

CHEM 332L Experiment 3: Aldol Condensation Reaction

Observations

1. Describe the acetone and benzaldehyde.

4. Describe any odors you may detect from afar.

5. Describe any color changes in the test tube as the reaction progresses.

6. Describe any apparent change in temperature of the test tube by the feel of your hand. When is there any bubbling?

Own Observations:

_____
_____
_____
_____
_____
_____
_____
_____
_____
_____
_____
_____
_____
_____
_____

RESULTS

Mass of acetone = _____ g

Mass of benzaldehyde = _____ g

Theoretical Yield of dibenzalacetone = _____ g

Percent Yield = _____ %

Percent Recovery = _____ %

Melting Point = _____ °C

Percent Difference = _____ %

CALCULATIONS

Conclusions

Experiment Four 4

Sodium Borohydride Reduction

OBJECTIVES

1. Perform the reduction of benzil to hydrobenzoin using sodium borohydride.
2. Isolate the product hydrobenzoin.
3. Calculate the theoretical yield.
4. Record the mass, obtain the melting point, and determine the percent yield.

THEORY

Reduction is the gain of electrons. A **reducing agent** is a compound that performs reduction and in return becomes **oxidized**. Conversely an **oxidizing agent** is a compound that performs an oxidation and becomes **reduced**.

1. Utility of borohydride reductions

Reducing Agents

Sodium Borohydride ($Na^+BH_4^-$)	Lithium Aluminum Hydride ($Li^+AlH_4^-$)
reduces ketones, and aldehydes	reduces carboxylic acids, epoxides, lactones (cyclic esters), nitro groups, nitriles, azides, amides, acid chlorides, esters, aldehydes, ketones
can be carried out in H_2O and alcohol solutions (protic solvents) [mostly carried out in ethanol (C_2H_5OH) or methanol (CH_3OH)]	mostly carried out in diethyl ether or tetrahydrofuran (aprotic solvents).
is insoluble in ether and soluble in methanol (CH_3OH) and ethanol (C_2H_5OH)	is soluble in ether and insoluble in methanol (CH_3OH) and ethanol (C_2H_5OH)
versatile: water, alcohols	reacts violently with H_2O and alcohols to evolve H_2 gas, forms metal hydroxides
mild and selective reducing agent	not a very selective reducing agent

2. Other Reducing Agents

Other Reducing agents include DiBal-H, which utilizes the hydride anion to reduce esters and nitriles to aldehydes. Catalytic reductions using hydrogen gas and a catalyst such as Platinum or Palladium will reduce C=C double bonds, triple bonds, ketones, Aldehydes, nitro-groups, amides and Imines.

3. Stereoisomers

Addition of four atoms of hydrogen to benzil gives a mixture of stereoisomeric diols, of which the predominant isomer is the nonresolvable (1R,2S)-hydrobenzoin, the meso isomer, accompanied by the enantiomeric (1R,2R) and (1S,2S) compounds. The intermediate borate ester is hydrolyzed with water to give the product alcohol. Benzil is used in this experiment rather than benzoin, because benzil is yellow and we can follow the reduction reaction.

    a. **Compatible with alcoholic solvents**
    b. **Selectivity**
    c. **Product formed (alcohol's)**

Benzoin

4. Reactions of borohydrides
    a. **Overall reaction**

Benzil (yellow solid)

1. $Na^+BH_4^-/C_2H_5OH$ (ethanol, polar protic solvent; reducing agent)
2. H_2O heat

(1R,2S)-(meso)-Hydrobenzoin (predominant isomer)

+ (1R,2R) and (1S,2S)-Hydrobenzoin (enantiomers)

+ $NaB(OH)_4$

CHEM 332L Experiment 4: Sodium Borohydride Reduction

5. Differences in melting point: Thought Question

Different isomers of the hydrobenzoin product have different melting points. The predominant isomer has a melting point range of **137-139 °C**. The enantiomer mixture is **120°C**. The question has been asked how do we separate the mixture of these two? A mixture of enatiomers will rotate plane-polarized light only if there is enatiomeric excess. Racemic mixtures will not rotate plane- polarized light. The same is true for the meso-compound which would also not rotate plane-polarized light. *So how do we separate the mixture that is Racemic?*

Since the functional group in question is an alcohol, we would resolve the mixture by forming diastereomeric salts. To resolve a racemic mixture of alcohols we use a chiral acid chloride. We can form diastereomeric esters, which have different specific rotations and different melting points. From this, we can then remove the ester that is formed and have clear R and S conformations from our Racemic Mixture.

CHEM 332L Experiment 4: Sodium Borohydride Reduction

EXPERIMENTAL FLOW DIAGRAM OF SEPARATION OF RACEMIC MIXTURE

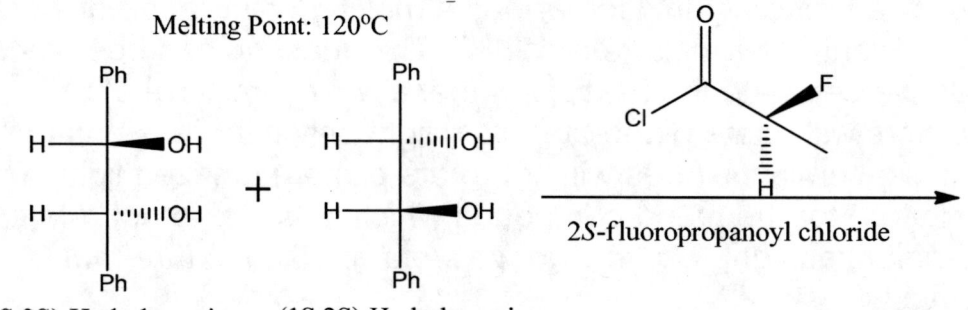

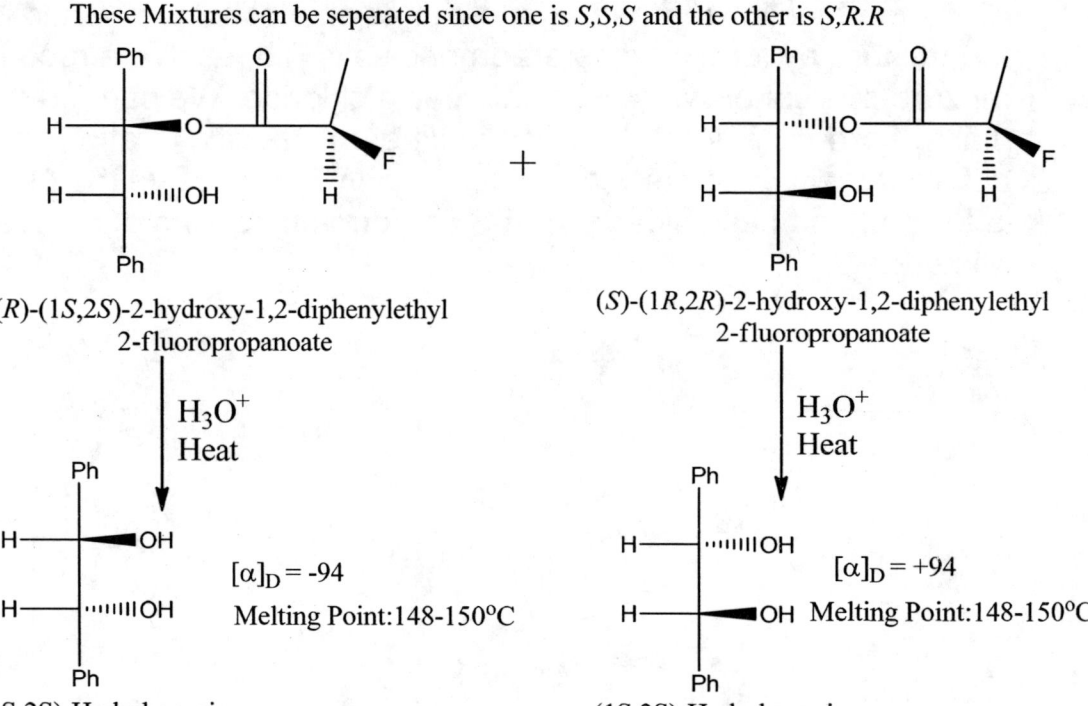

236

CHEM 332L Experiment 4: Sodium Borohydride Reduction

Reaction Mechanism

Benzil + Sodium Borohydride →

→ Dialkoxy Boron Complex + Benzil

→ Tetra Alkoxy Boron Complex (borate ester) $\xrightarrow[4\ H_2O]{heat}$ 2 (Hydrobenzoin mixture) + $NaB(OH)_4$

REAGENTS

Name, Structure, MW (g/mol)	Melting °C	Boiling °C	Density g/mL	Properties	Safety
Benzil $C_{14}H_{10}O_2$ 210.2318	95	346 – 348	-	Fine yellowish powder	Causes eye irritation; may cause chemical conjunctivitis; Causes skin irritation; may cause irritation of the digestive tract; may cause gastrointestinal irritation with nausea, vomiting and diarrhea; may cause respiratory tract irritation; can produce delayed pulmonary edema
Ethanol C_2H_5OH 32.042	-114.1	78.3	0.789	Colorless liquid; pleasant alcoholic odor	Harmful by ingestion, inhalation or skin absorption; irritant of the eyes, nose and throat, and skin; flashbacks along the haphazard vapor trails; at decomposition, emits toxic fumes of CO and CO_2
sodium borohydride $NaBH_4$ 37.83137	36	400	.945	Hydroscopic, cubic crystals; reducing agent for the conversion of aldehydes and ketones to alcohols; HYGROSCOPIC; DANGEROUS WHEN WET	Dust may cause severe irritation to respiratory system; causes severe irritation and burns to the digestive tract; severe irritation and skin burns may result when the chemical comes in contact with skin; if contact occurs, then no immediate warning will occur until burning or irritation begins
hydrobenzoin $C_{14}H_{14}O_2$ 214.2634	138	-	-	-	-

CHEM 332L Experiment 4: Sodium Borohydride Reduction

SAFETY

- ✓ As a heating mantle is in use, extra precaution must be taken.
- ✓ The apparatus can also get quite hot during use.
- ✓ Take care to avoid broken glass and sharp objects at all times.
- ✓ Gloves will be provided when you enter the lab in order to prevent direct contact of the organic solvents with your skin. Wear goggles at <u>all</u> times.
- ✓ Sodium borohydride is extremely corrosive.

DISPOSAL

- ✓ Wash excess aqueous solutions down the drain with excess water.
- ✓ Organic solvents are to be placed in predesignated organic liquid waste container separate from other organic liquid waste.
- ✓ Dispose of soiled gloves and paper towels in trash. Dispose of broken glass in the *broken glass container*.

CHEM 332L Experiment 4: Sodium Borohydride Reduction

PROCEDURE

REMINDER: Close and cap all reagent and waste containers.

1. Dissolve 50 mg of benzil in 0.5 mL of 95% ethanol (C_2H_5OH) in a small test tube.

2. Cool the solution in an ice bath. A fine yellow suspension should form.

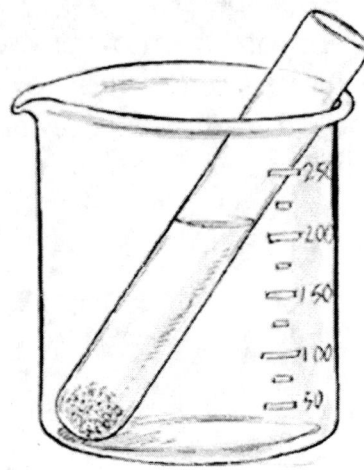

3. Carefully add 10 mg (.01g) of sodium borohydride all at once to the test tube. Make note of the following:
    a. The time the solid takes to go into solution.
    b. Evolution of heat during the course of the reaction. *Is this an endothermic or exothermic reaction?*
    c. Time required for the yellow color to disappear (3 - 4 minutes).

CHEM 332L Experiment 4: Sodium Borohydride Reduction

4. Let the mixture stand for 10-12 minutes with occasional swirling. This allows the reduction to go to completion.

5. Add 0.5 mL of water to the tube and heat the reaction mixture to boiling in order to hydrolyze the borate ester to the corresponding secondary alcohols.

CHEM 332L Experiment 4: Sodium Borohydride Reduction

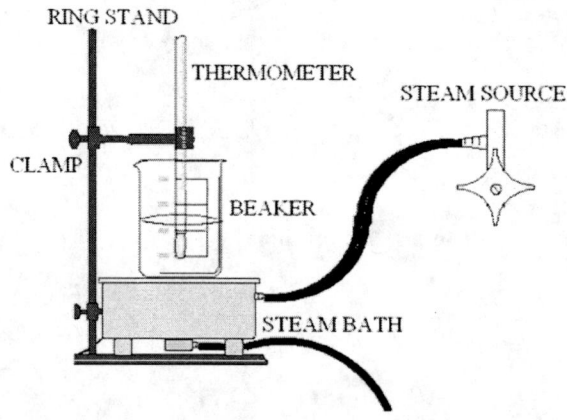

6. Remove the test tube from the hot bath and add 1 mL of hot water. A cloudy mixture will form. The product, hydrobenzoin, will begin crystallizing as thin plates.

7. Complete the crystallization process by placing the reaction mixture in an ice bath.

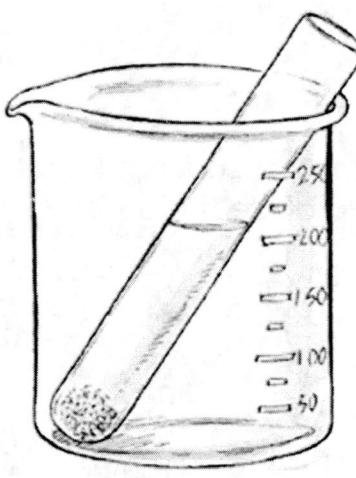

CHEM 332L Experiment 4: Sodium Borohydride Reduction

8. Collect the product by suction filtration. Wash the product with ice cold water.

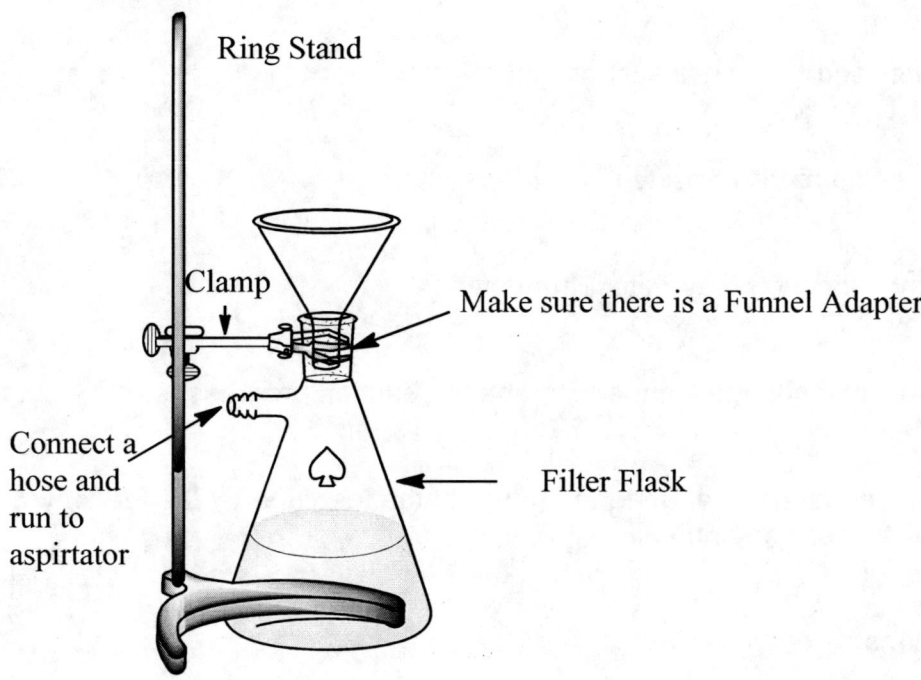

9. Air dry using suction filtration for 10 minutes.

10. Record the mass of the product, determine the melting point, and calculate the percent yield.

$$\% \text{ Yield} = 100 \times \frac{\text{Actual Yield}}{\text{Theoretical Yield}}$$

CHEM 332L Experiment 4: Sodium Borohydride Reduction

OBSERVATIONS

1. Describe benzil and the benzil suspension.

2. Describe the sodium borohydride.

3. Describe any odors you may detect from afar.

4. Describe any color changes in the test tube as reaction progresses.

5. Describe any apparent change temperature of the test tube by the feel of your hand. When is there any bubbling?

Own Observations:

_____
_____
_____
_____
_____
_____
_____
_____
_____
_____
_____
_____
_____
_____
_____
_____

RESULTS

Mass of benzil = _____ g

Theoretical Yield of product hydrobenzoin = _____ g

Mass of product hydrobenzoin = _____ g

Percent Yield = _____ %

Melting Point = _____ °C

Percent Difference = _____ %

CALCULATIONS

CONCLUSIONS

Experiment Five

Acid Catalyzed Acylation

OBJECTIVES
1. Synthesize an ester from an unknown alcohol using acid catalyzed acylation.
2. Isolate and purify the product using a distillation apparatus.
3. Determine the boiling point.

THEORY

A.) Properties of Esters:
Low molecular weight esters have very pleasant odors and are the major components of the flavor and odor of many fruits. Esters are used in the food industry as artificial flavors and in the perfume industry as fragrances. Esters can be prepared through the reaction of a carboxylic acid with an alcohol in the presence of an acid. For our experiment we will be using acetic anhydride and an alcohol to form our ester.

B.) Acid Catalysis:
Fischer Esterification has been a popular choice for organic labs for many years. However, due to low yields and time constraints, it has not proven to be a very productive way to synthesize an ester. By using acetic anhydride (a very reactive compound) instead of acetic acid, the synthesis of an ester can be accelerated. As with the Fischer Esterification reaction, the acetic anhydride reaction is catalyzed by H_2SO_4. Since this reaction is a very exothermic reaction, the first part of this experiment is carried out in an ice bath followed by heating on a steam bath to complete the reaction.

n-propanol + acetic anhydride $\xrightarrow{H_3O^+ \text{ Catalyst}}$ acetic acid + propyl acetate

C.) Reversibility and the Role of Sulfuric Acid:

The Fischer esterification reaction will reach equilibrium after a few hours of heating under reflux and the equilibrium can be shifted more towards the product with the addition of either more alcohol or more catalyst. This process is completely reversible. Unlike a Fischer Esterifcation reaction, using Acetic Anhydride in the ester synthesis is fast and irreversible.

Alternate Methods of Esterification:

1.) Fischer Esterfication. Refluxing the appropriate carboxylic acid with the appropriate alcohol for several hours will result in the ester of choice.

1-butanol + acetic acid $\xrightleftharpoons[\text{Heat}]{H_2SO_4}$ butyl acetate

2.) Reacting an alcohol with an acid chloride gives an ester and hydrochloric acid is a byproduct.

ethanol + acetyl chloride $\rightleftharpoons$ ethyl acetate + HCl

3.) Reaction of diazomethane (in a solution of diethyl ether) gives methyl esters, with the evolution of nitrogen gas.

$[\ ^-N\!=\!N^+\!=\!CH_2$ diazomethane $\leftrightarrow \ddot{N}\!\equiv\!N^+\!-\!CH_2^-]$

+ acetic acid $\xrightarrow{\ \ \ }$ methyl acetate + N_2

D.) Pressure build up

Pressure build up in an extraction procedure that contains a very volatile solvent can be caused from either the heat from acid/base reactions or even the warmth of the hand. This means that whatever apparatus is being used, it must be vented frequently (ex. Separatory funnel). Sodium bicarbonate is used to neutralize acids when doing an acid/base extraction. Upon addition of the sodium bicarbonate, carbon dioxide will form and will cause a pressure build up.

$$H_3O^+ + NaHCO_3 \rightarrow HCO_3^- + Na^+$$

$$HCO_3^-(aq) + H^+(aq) \rightarrow H_2CO_3(aq)$$

$$H_2CO_3(aq) \rightarrow H_2O(l) + CO_2(g)$$

CHEM 332L Experiment 5: Acid Catalyzed Acylation

MECHANISM: ACYLATION OF AN ALCOHOL

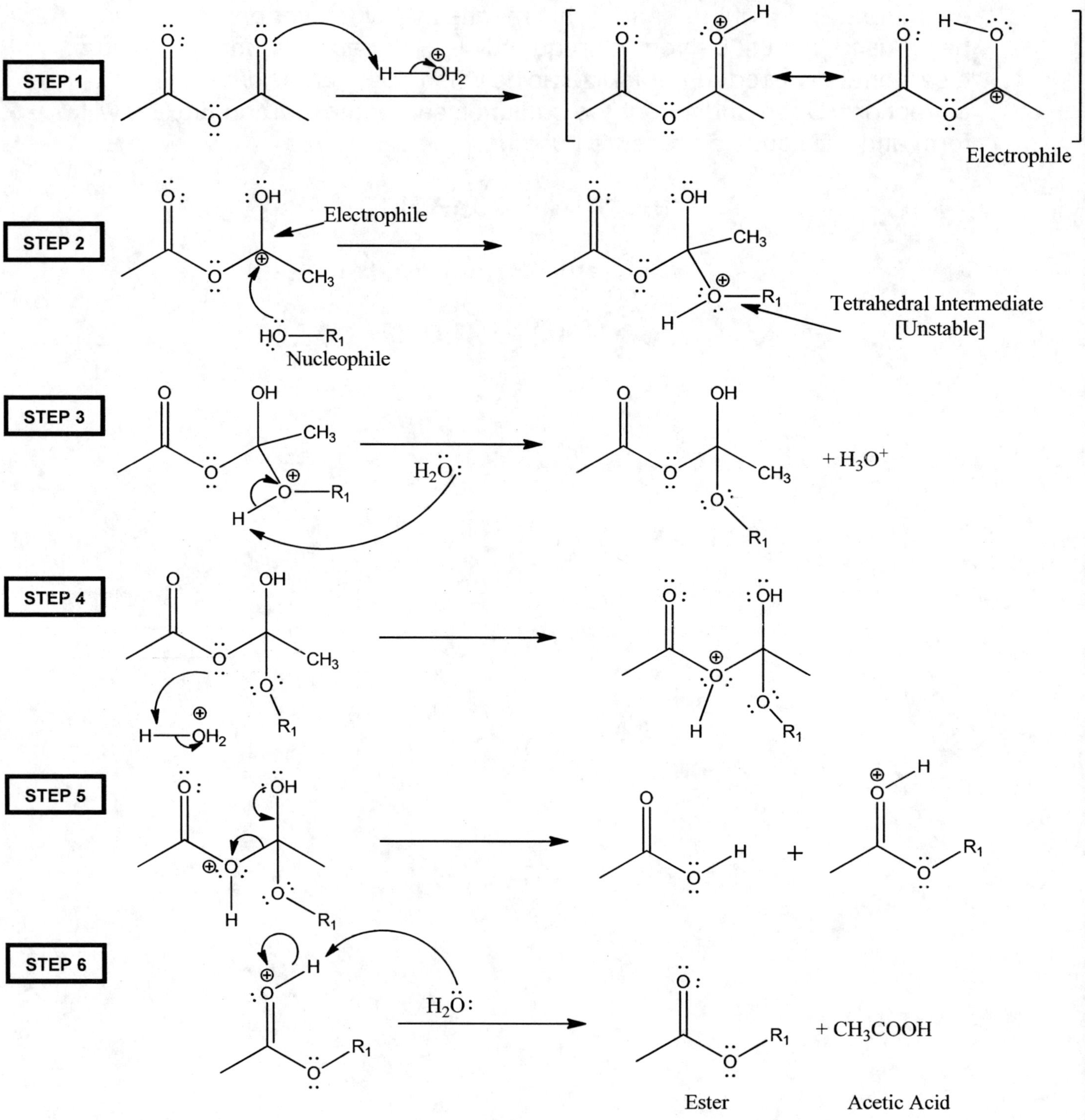

STEP 1

STEP 2

STEP 3

STEP 4

STEP 5

STEP 6

REAGENTS

Name, Structure, MW (g/mol)	Melting Pt. °C	Boiling Pt. °C	Density (g/mL)	Properties	Safety
Acetic Acid $C_2H_4O_2$ 60.05	16.6	118	1.05	Corrosive material; colorless liquid with a strong, vinegar-like odor	Corrosive to the skin, eyes, and mucous membranes
Diethyl Ether $C_4H_{10}O$ 74.1224	-116	34.6	.706	Colorless liquid with a characteristic, sweet, ether odor	Moderately toxic to humans by ingestion; absorbs through skin; FLAMMABLE
Magnesium sulfate $MgSO_4$ 120.3626	1124	-	2.66	HYGROSCOPIC; Drying agent; White powder	Dessicant
Sodium bicarbonate $NaHCO_3$ 84.001	801	1413	1.03	White crystals	When heated to decomposition, this compound emits toxic fumes of carbon monoxide, carbon dioxide and sodium oxides
Sulfuric Acid H_2SO_4 98.07	3	290	1.84	Corrosive material; Colorless (pure) to dark brown; Oily, dense liquid with sharp acrid odor; HYGROSCOPIC	Highly corrosive; causes severe, deep burns on eye and skin contact and upon inhalation of sulfuric acid mist; highly reactive, reacts violently with many organic and inorganic substances
Sodium chloride NaCl 58.442	801	1413	2.165	White crystals; HYGROSCOPIC	May cause eye irritation; when heated to decomposition it emits toxic fumes

CHEM 332L Experiment 5: Acid Catalyzed Acylation

Name, Structure, MW	mp (°C)	bp (°C)	density	Appearance	Hazards
1-Butanol $C_4H_{10}O$ 74.12	-90	111	0.81	Clear liquid	May cause mild irritation of the nose, throat and respiratory tract
1-Propanol C_3H_7OH 60.11	-126.5	97	0.804	Colorless liquid with a mild, non-residual, alcoholic odor; HYGROSCOPIC	Mild irritant of the eyes, skin, nose, and lungs
Isopropanol C_3H_7OH 60.11	-89.5	82	0.785	Clear colorless liquid with an odor of rubbing alcohol; HYGROSCOPIC	Mild irritant of the eyes, skin, nose, and lungs
n-Hexanol $C_6H_{13}OH$ 102.18	-52	157	0.814	Colorless liquid	Mild irritant of the eyes, skin, nose, and lungs
1-Octanol $C_8H_{18}O$ 130.14	-15	196	0.827	Colorless liquid with a penetrating odor	Mild irritant

BOILING POINTS OF THE ESTERS

Name, Structure, MW (g/mol)	Boiling Pt. °C	Properties
Methyl Acetate $C_3H_6O_2$ 74.08	57	Smells like many nail polishes, and model airplane glue.
Propyl Acetate $C_5H_{10}O_2$ 102.1	102	Odor of pears. Common Flavoring Compound.
Isopropyl Acetate $C_5H_{10}O_2$ 102.1	89	Common Odor in printer inks, characteristic fruity odor. Not specific fruit smell however.
Butyl Acetate $C_6H_{12}O_2$ 116.16	125	Smell of bananas. Gives Red delicious apples their characteristic flavor.
Hexyl Acetate $C_8H_{16}O_2$ 144.21	169	Green fruity smell slightly reminiscent of pears. Sweet smell similar to benzaldehyde.
Octyl Acetate $C_{10}H_{20}O_2$ 170.27	211	Basis for artificial orange flavoring. Orange smell.

CHEM 332L Experiment 5: Acid Catalyzed Acylation

SAFETY
- Wash excess aqueous solutions down the drain with excess water.
- Gloves will be provided when you enter the lab in order to prevent direct contact of the organic solvents with your skin. Wear goggles at <u>all</u> times.
- Organic solvents (ether) are to be placed in pre-designated organic liquid waste container separate from other organic liquid waste.
- Dispose of soiled gloves and paper towels in proper containers. **Dispose of broken glass in the *broken glass container*.**
- All Esters synthesized go in the organic liquid waste container.

DISPOSAL
- **Be Sure that your test tube is completely submerged in the ice bath when you add the alcohol to your reaction!!!**
- <u>**DO NOT**</u> add boiling chips to a hot liquid. A superheated liquid may boil over or even explode.
- As a hot heating mantle is in use, extra precaution must be taken.
- The apparatus can also get quite hot during use.
- Take care to avoid broken glass and sharp objects at all times.
- As with all acids and bases, extra precaution should be taken when sulfuric acid is in use.
- There are flammable and volatile liquids in use.
- Gloves will be provided when you enter the lab in order to prevent direct contact of the organic solvents with your skin. Wear goggles at <u>all</u> times.

PROCEDURE
REMINDER: Close and cap all reagent and waste containers.
1. Set up a large beaker containing water on the steam bath at your work station placed on the steam bath. In another large beaker, set up an ice bath filled with ice and water.

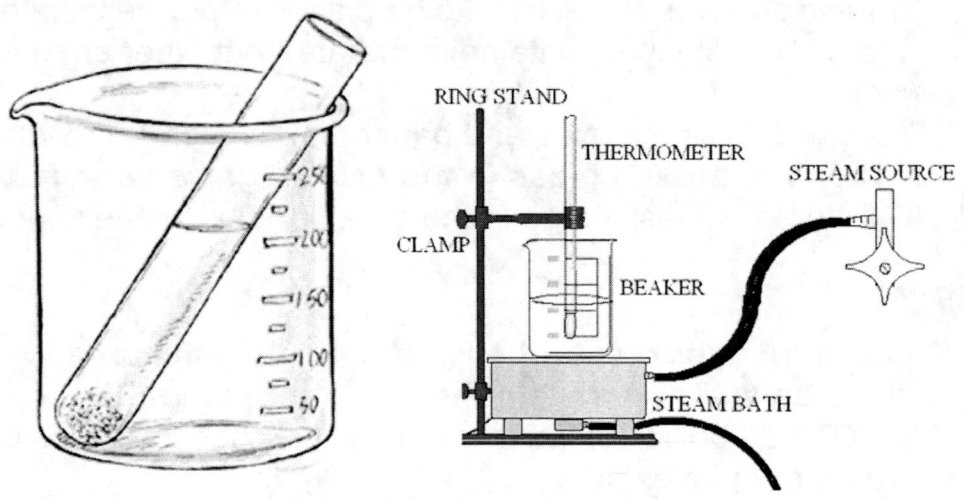

2. Place 2mL of acetic anhydride in a *large, dry* test tube. Add 4 drops of concentrated sulfuric acid and mix thoroughly.

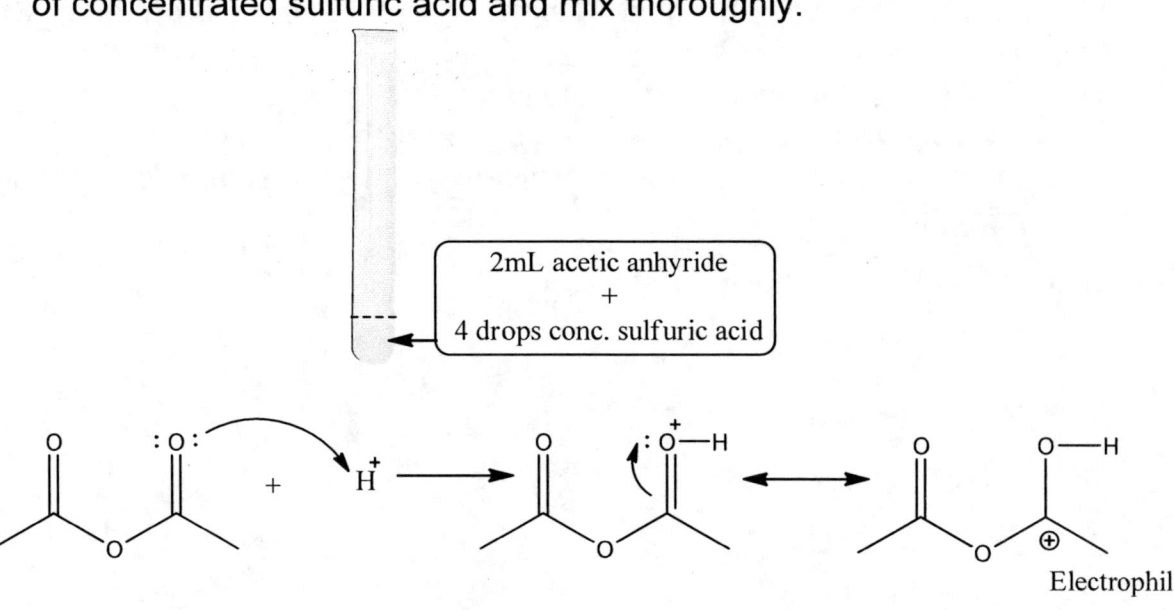

CHEM 332L Experiment 5: Acid Catalyzed Acylation

3. Obtain 2.0 mL of one of the alcohols in a dry small test tube.

4. Hold the large test tube containing the acetic anhydride in the ice bath (to cool the reaction). Add the alcohol dropwise using a Pasteur pipette. Mix between additions. Cooling is therefore necessary. The strongly exothermic reaction could become too vigorous.

5. After the addition of the alcohol is complete, place the tube in the hot water bath at about 70° C for 7 min to complete the reaction.

(Ester)   (Acetic Acid)

6. When the reaction has gone to completion, add eight drops of water using a Pasteur pipet. Mix briefly after each drop. This is done to hydrolyze any excess un-reacted acetic anhydride remaining after the esterification is complete.

7. Cool the reaction mixture to nearly room temperature and add 6 mL half saturated sodium chloride solution (3 mL saturated NaCl plus 3 mL distilled water) to the test tube. Mix thoroughly and vigorously, then set the tube aside until upper and lower layers form. **Allow the test tube to sit for 3 minutes to allow for the separation of layers completely.** The upper layer is the <u>ester</u>, which is a water-insoluble volatile liquid.

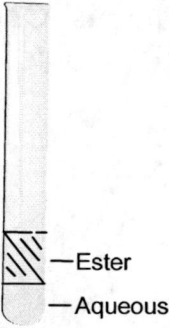

8. Use a Pasteur pipet to remove the lower aqueous layer. The last portion of the lower layer can be removed by placing the tip of the Pasteur pipet against the bottom of the tube, and observing carefully as you slowly withdraw the liquid. Do not be concerned if a few drops remain behind. Do not remove the upper layer (the ester) from the test tube.

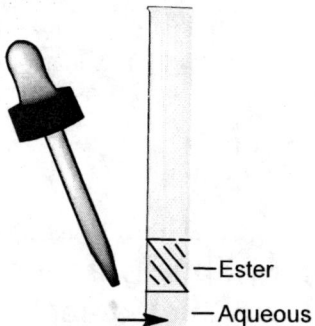

CHEM 332L Experiment 5: Acid Catalyzed Acylation

9. Add 6.0 mL of saturated sodium bicarbonate to the upper layer in the test tube, mix vigorously, and allow up to 5 minutes for the layers to separate. The base neutralizes remaining acids (eq 3 and 4), and helps to remove traces of acetic acid. Remove and discard the lower layer. <u>Retain the upper layer.</u>

$$CH_3COOH + NaHCO_3 \longrightarrow CH_3COO^- Na^+ + H_2CO_3$$
$$H_2CO_3 \longrightarrow CO_2 + H_2O$$

10. Again wash the upper layer ether layer with 6.0 mL of half-saturated NaCl (Repeat steps # 8 and 9).

11. Wash the upper layer with 6.0 mL saturated sodium chloride. Mix thoroughly, and allow the layers to separate. The upper layer (the ester) should not be cloudy at this point. Again remove and discard the lower layer, but this time try to remove every drop of it (without, of course, removing a significant amount of the ester).

12. Transfer the ester to a clean dry test tube and dry the ester by adding a micro-scoop of MgSO$_4$. Allow the magnesium sulfate to settle in the test tube for 3-5 miutes to allow all the water to be removed from the sample.

$$R-C(=O)-O-CH_3 \cdot XH_2O \xrightarrow{MgSO_4} MgSO_4 \cdot XH_2O + Ester$$

CHEM 332L Experiment 5: Acid Catalyzed Acylation

13. Add the dried ester into to a separate clean dry test tube and record the weight of the ester. Record the mass.

14. Perform a micro-boiling point determination of your final ester to determine the boiling point and identify the ester synthesized.

CHEM 332L Experiment 5: Acid Catalyzed Acylation

OBSERVATIONS

1. Describe the unknown alcohol.

2. Describe any odors you may detect from afar. DO NOT INHALE VAPORS.

3. Describe any color changes in the round bottom flask as reaction progresses.

4. Describe the product.

Own Observations:

_____
_____
_____
_____
_____
_____
_____
_____
_____
_____
_____
_____
_____
_____
_____

RESULTS

Amount of Product = _____ g

Boiling Point = _____ °C

Identity of Unknown Alcohol = _____

CALCULATIONS

CONCLUSIONS

Experiment Six

Diels-Alder Reaction

OBJECTIVES

1. Perform a Diels-Alder reaction between anthracene and maleic anhydride.
2. Isolate the product using suction filtration.
3. Calculate the theoretical yield.
3. Record the mass and determine the melting point.

THEORY

1. Reactants

The Diels-Alder reaction is a concerted cycloaddition reaction between a diene and a dienophile.

Diene   Dienophile

2. General Mechanism

The prototype Diels-Alder reaction is the *cycloaddition* of butadiene (diene) and ethylene (dienophile) gases.

Under high pressure, the gas molecules are forced together and undergo a concerted reaction involving the simultaneous movement of six electrons.

In cases when the R groups coming off the dienophile are electron withdrawing (-COOH, -COH, -CN), the overall reaction rate of the Diels-Alder will be increased. Conversely, electron donating groups (-OH, -NH$_2$) tend to slow down the rate of the reaction.

3. Endo vs. Exo Product

Since the Diels-alder is a concerted reaction, several things happen to the addition, especially when a bi-cyclic structure is formed (as in the case of this experiment). If the dienophile's substituents point away from the diene, the product will be an exo-product. When the substituents of the dienophile point towards the diene, the endo-product is formed. In other words, when looking from a side angle, it can be perceived that the exo-product is outside relative to the double bond, and the endo-product is inside the double bond. Typically, the endo product is more sterically hindered, so it would seem logical that the exo is more readily formed. However, the endo product is the predominant isomer due to systems in which the substituent coming off the dieneophile can participate in the conjugation with that double bond. In our case, the product is an endo since maleic anhydride does have a π bond system that is in conjugation.

4. Overall reaction

9,10-dihydroanthracene-
9,10-alpha,beta-
succinic acid anhydride

The reactants in this experiment are anthracene and maleic anhydride. Which compound is the diene, and which is the dienophile?

5. Mechanism:

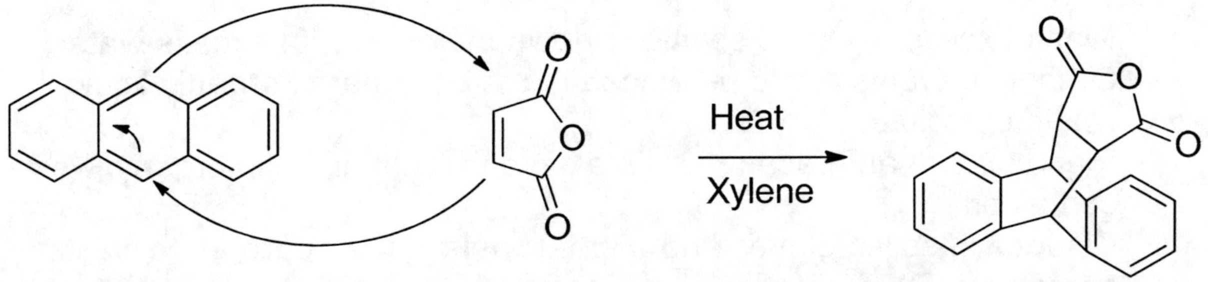

REAGENTS

Name, Structure, MW (g/mol)	Melting Pt. °C	Boiling Pt. °C	Density (g/mL)	Properties	Safety
anthracene 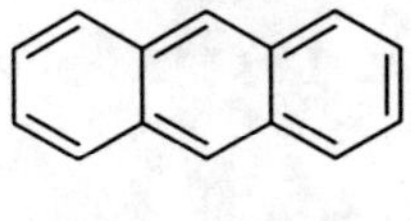 $C_{14}H_{10}$ 178.23	213	340	1.25	Colorless solid	Can cause irritation
maleic anhydride 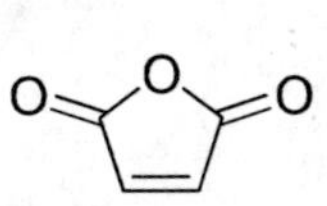 $C_4H_2O_3$ 98.06	51-56	200	1.43	Colorless liquid with a characteristic, sweet, ether odor	Extreme eye irritant; can cause dermatitis
p-xylene 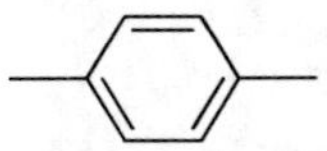 C_8H_{10} 120.3626	12.5	138	0.861	HYGROSCOPIC; Drying agent; White powder	Dessicant

CHEM 332L Experiment 6: Diels-Alder Reaction

SAFETY
- ✓ Wash excess aqueous solutions down the drain with excess water.
- ✓ Organic solvents are to be placed in pre-designated organic liquid waste container.
- ✓ Place excess anthracene and maleic anhydride in solid organic waste container.
- ✓ Dispose of soiled gloves and paper towels in the appropriate waste containers. Dispose of broken glass in the *broken glass container*.
- ✓ The product goes in the organic solid waste container.

DISPOSAL
- ✓ Make sure that the reflux apparatus is tightly sealed.
- ✓ **DO NOT** add boiling chips to a hot liquid. A superheated liquid may boil over or even explode.
- ✓ As a hot heating mantle is in use, extra precaution must be taken.
- ✓ The apparatus can also get quite hot during use.
- ✓ Take care to avoid broken glass and sharp objects at all times.
- ✓ Gloves will be provided when you enter the lab in order to prevent direct contact of the organic solvents with your skin. Wear goggles at <u>all</u> times.
- ✓ There are flammable and volatile liquids in use.

PROCEDURE

REMINDER: Close and cap all reagent and waste containers.

NOTE: High temperatures are needed for this reaction, so it is a good idea to begin heating your heating mantle as soon as you enter the lab!

1. Weigh and place 0.80 g of anthracene and 0.40 g of maleic anhydride in a clean, dry round-bottom flask.

   0.8 grams Anthracene   +   0.4 grams maleic anhydride

2. Add 10 mL of *p*-xylene to the flask.

   p-xylene [Solvent]

3. Attach a reflux condenser. Clamp down the apparatus securely.

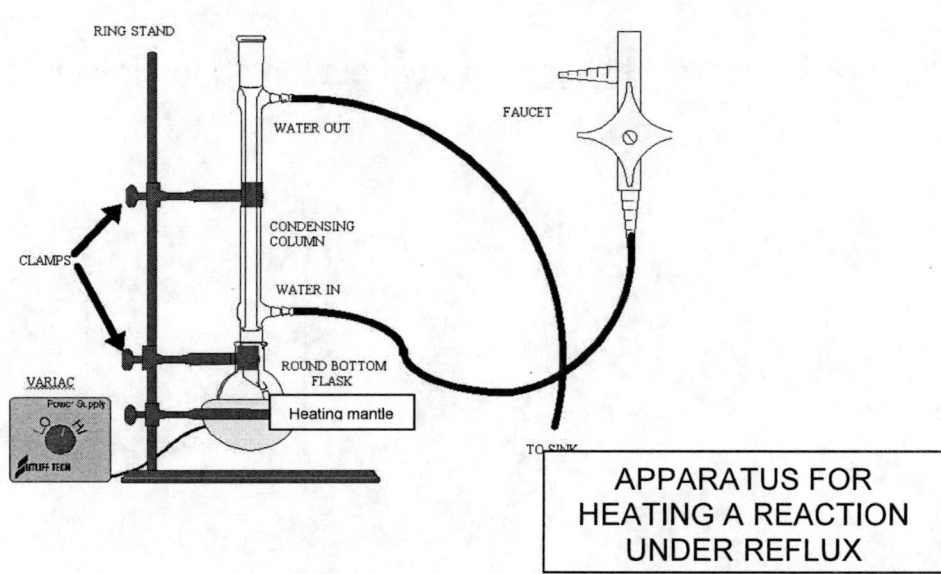

APPARATUS FOR HEATING A REACTION UNDER REFLUX

CHEM 332L Experiment 6: Diels-Alder Reaction

4. Heat the reaction mixture on a heating mantle to reflux temperature (185-200°C).

5. Heat under reflux the mixture for 30 minutes. <u>Note any color changes in Observations.</u>

[Diene] + [Dienophile] →(Heat, Xylene)→ product

6. Remove the heat and allow the reaction mixture to cool to room temperature.

9,10-dihydroanthracene-9,10-alpha,beta-succinic acid anhydride

7. Place 3 mL of <u>ethyl acetate</u> in a test tube and place it in an ice bath.

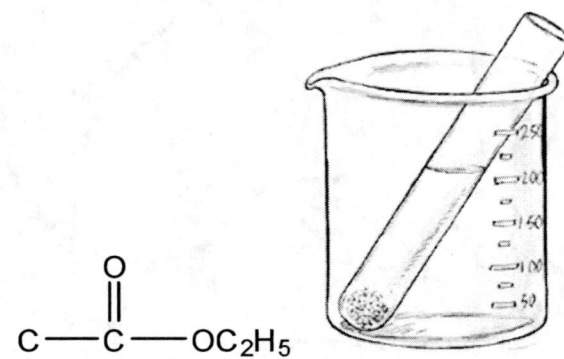

268

8. Once the flask has reached room temperature, place it in the ice bath for 10 minutes to complete **[At Least]** the crystallization.

9. Collect the crystals using suction filtration, washing the product with **COLD** ethyl acetate.

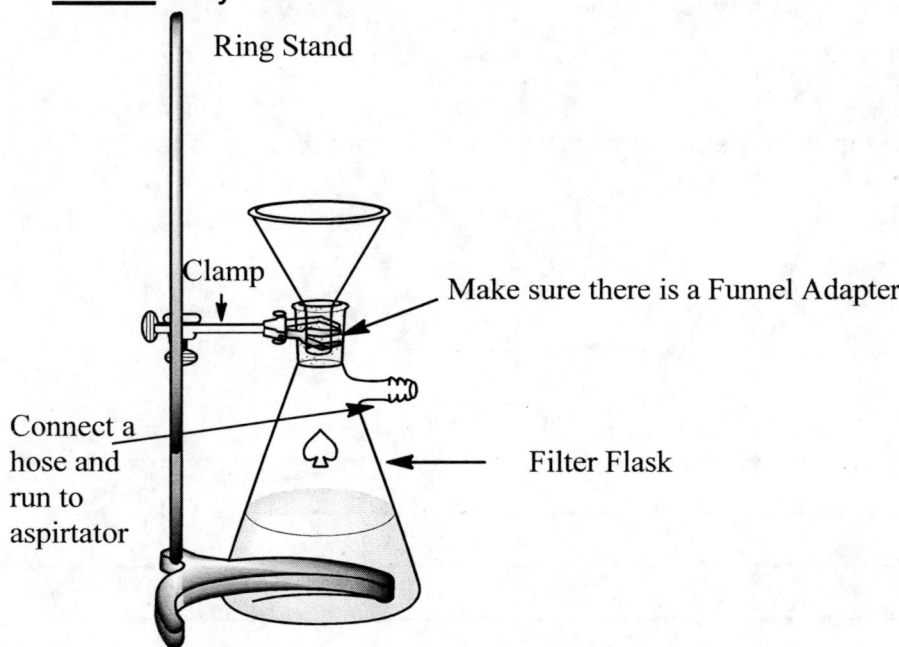

10. Air dry the product with suction filtration for 5 minutes.

11. Calculate the theoretical yield. What is the limiting reagent?

12. Record the mass of the product, determine the melting point, and calculate the percent yield.

$$\% \text{ Yield} = 100 \times \frac{\text{Actual Yield}}{\text{Theoretical Yield}}$$

CHEM 332L Experiment 6: Diels-Alder Reaction

OBSERVATIONS

1. Describe the starting materials.

2. Describe any odors you may detect from afar.

3. Describe any color changes in the round bottom flask as reaction progresses.

4. Describe the product.

Own Observations:

_____
_____
_____
_____
_____
_____
_____
_____
_____
_____
_____
_____
_____
_____
_____

RESULTS

Mass of Product = _____ g

Melting Point = _____ °C

Percent yield = _____ %

CALCULATIONS

Conclusions

Experiment Seven 7

Solvent-Free Claisen Condensation of Ethyl Phenylacetate

OBJECTIVES

1. Set up a 25 mL round bottom flask equipped with a reflux condenser.
2. Synthesize ethyl 3-oxo-2,4-diphenylbutanoate from ethyl phenylacetate via a Claisen condensation reaction.
3. Record the mass, determine melting point, and calculate percent yield.

THEORY

ethyl 2-phenylacetate → (1. Potassium tert-butoxide, 2. HCl) → ethyl 3-oxo-2,4-diphenylbutanoate

Nucleophilic acyl substitution is a reaction in which a nucleophile bonded to the carbon of an acyl group is replaced by another nucleophile. Because aldehydes and ketones do not have a group that can leave as a relatively stable ion, they cannot undergo this type of substitution.

CHEM 332L Experiment 7: Claisen Condensation

1. **The Claisen Condensation:**
    The mechanism for the Claisen reaction is similar to that of the aldol reaction. The Claisen condensation involves the nucleophilic addition of an ester enolate ion donor to the carbonyl group of a second ester molecule.

    The only difference between the aldol condensation reaction and the Claisen condensation reaction is the tetrahedral intermediate in the aldol reaction is protonated to give an alcohol as seen previously for the aldol condensation reaction of ketones and aldehydes. The tetrahedral intermediate in the Claisen reaction eliminates an alkoxide leaving group to yield the acyl substitution product.

$$R_2N^- \quad RO^- \quad RCO_2^- \quad X^-$$

$$\xrightarrow{\text{Increasing leaving ability}}$$

Other Products:
From the Beta-Keto Ester, a Beta-Keto Acid can be formed by acid or base hydrolysis. In addition, a ketone could be formed by the thermal decarboxylation of the β-keto acid ketone. Shown below is a general reaction showing the different products that can be made from the Claisen condensation.

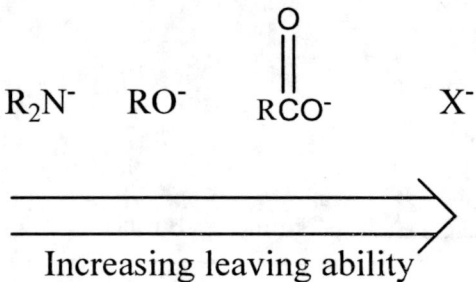

2. The Dieckmann Cyclization:

An intramolecular Claisen condensation can be done with acyclic diesters just as intramolecular aldol condensations can be carried out with acylic diketones. An intramolecular Claisen condensation is called a Dieckmann cyclization. The reaction works best on 1,6-diesters and 1,7-diesters. Five membered cyclic β–keto esters result from the Deickman cyclization of 1,6-diesters, and six membered cyclic β–keto esters result from cyclization of 1,7-diesters.

1.

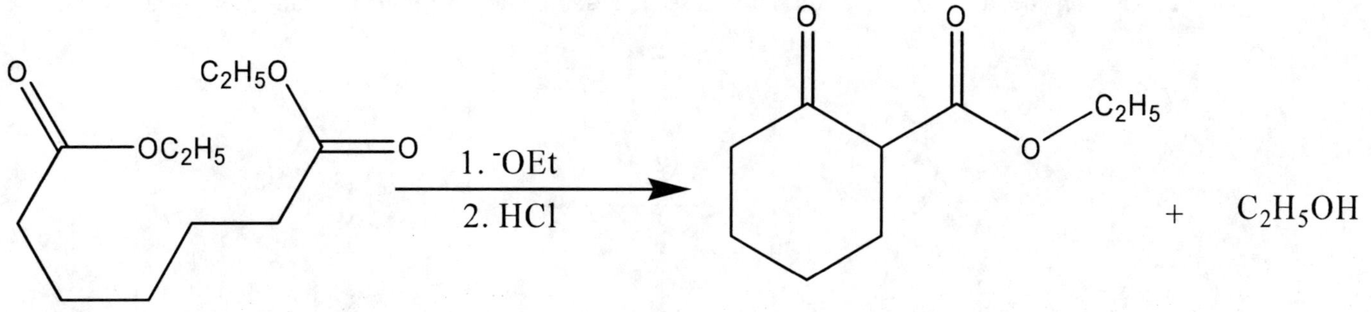

diethyl heptanedioate $\xrightarrow{\text{1. }^-\text{OEt} \quad \text{2. HCl}}$ ethyl 2-oxocyclohexanecarboxylate + C_2H_5OH

2.

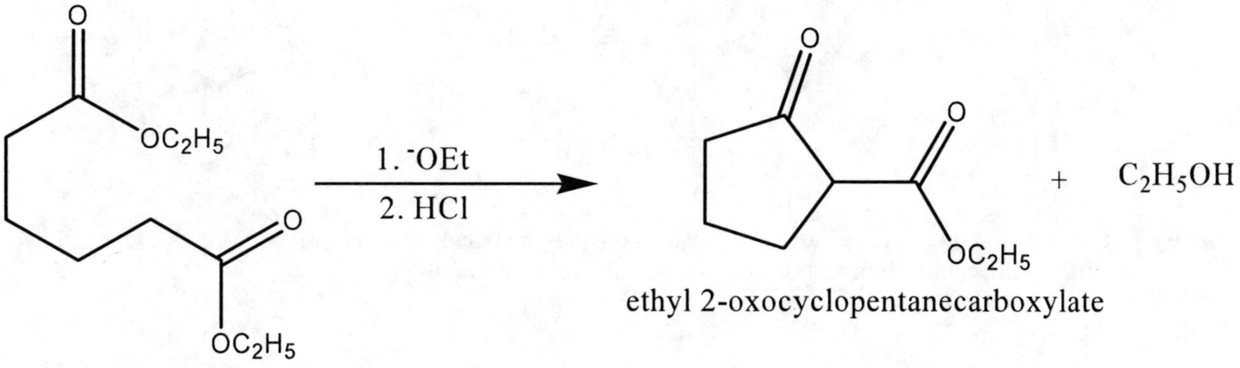

diethyl adipate $\xrightarrow{\text{1. }^-\text{OEt} \quad \text{2. HCl}}$ ethyl 2-oxocyclopentanecarboxylate + C_2H_5OH

CHEM 332L Experiment 7: Claisen Condensation

Stepwise Mechanism:

Step 1 - Potassium *tert*-butoxide extracts an acidic alpha hydrogen from an ester molecule, yielding an ester enolate

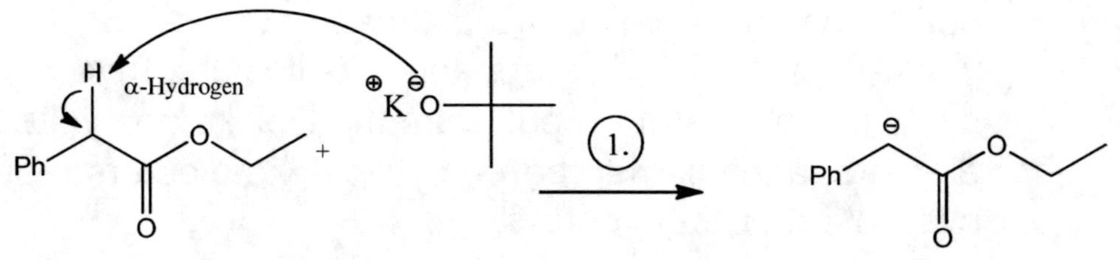

Step 2 - This ion does a nucleophilic addition to a second ester molecule, giving a tetrahedral intermediate.

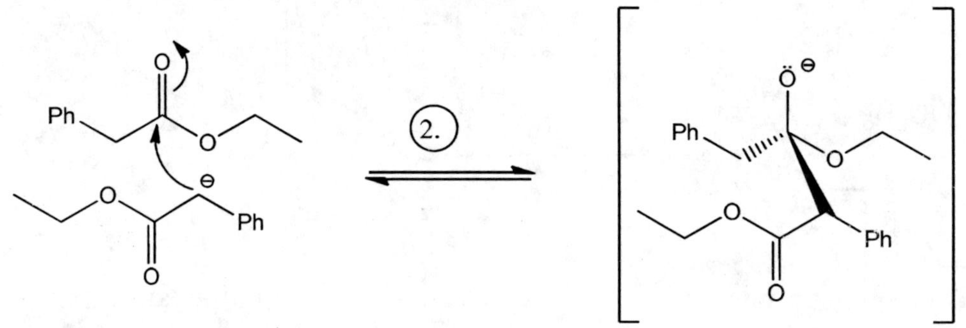

Tetrahedral Intermediate

Step 3 - The tetrahedral intermediate is not stable. It expels ethoxide ion to yield the new carbonyl compound.

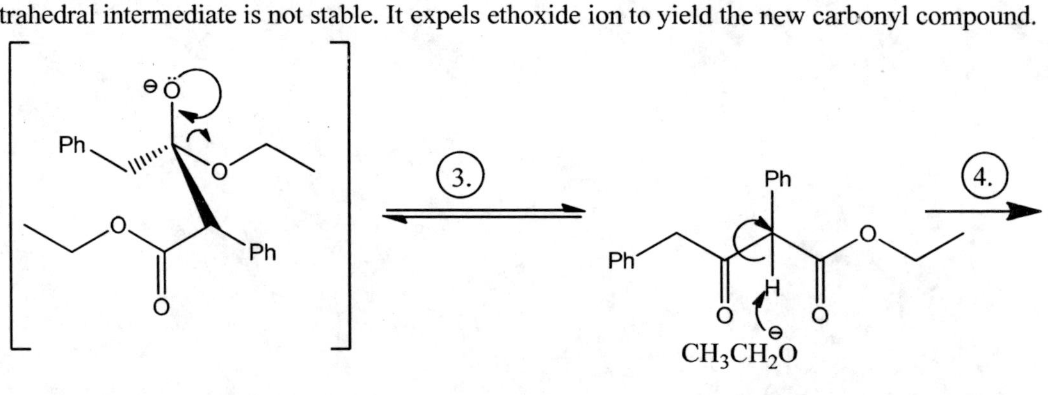

Step 4 - But, the ethoxide ion is a base. It therefore converts the beta keto ester product into its enolate, thus shifting the equilibrium and driving the reaction to completion.

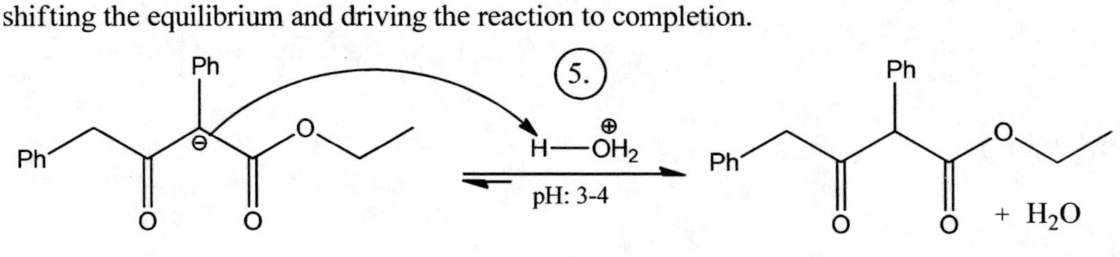

Step 5 - Protonation by addition of acid yields the final product.

REAGENTS

Name, Structure, MW (g/mol)	Melting Pt. °C	Boiling Pt. °C	Density (g/mL)	Properties	Safety
Potassium *tert*-butoxide C_4H_9KO 112.22	257	NA	-	White crystalline powder	Corrosive to the skin, eyes, and mucous membranes; Reacts violently with water
Diethyl Ether $C_4H_{10}O$ 74.12	-116.3	34.6	.706	Colorless liquid with a characteristic, sweet, ether odor	Moderately toxic to humans by ingestion; absorbs through skin; FLAMMABLE
Pentane C_5H_{12} 72.15	-130	36	.626	Clear, colorless solution	Irritant of the eyes, skin and respiratory tract; mild narcotic in high concentrations
Magnesium sulfate $MgSO_4$ 120.37	1124	-	2.66	HYGROSCOPIC; Drying agent; White powder	Dessicant
Ethyl phenylacetate $C_{10}H_{12}O_2$ 164.20	NA	229	1.03	Liquid	Irritant
Hydrochloric Acid HCl 36.46	-30	100	1.2	Corrosive material	Highly corrosive irritant of the skin, eyes and mucous membranes; CORROSIVE
Hexane C_6H_{14} 86.1766	-95	69	.659	Colorless flammable liquid with a mild gasoline–like odor	Irritating to the eyes, skin and respiratory tract; narcotic in high concentrations; when heated, emits smoke and toxic fumes; FLAMMABLE

SAFETY

- ✓ **DO NOT** add boiling chips to a hot liquid. A superheated liquid may boil over or even explode.
- ✓ Make sure that the reflux apparatus is tightly sealed
- ✓ As a heating mantle is in use, extra precaution must be taken.
- ✓ The apparatus can also get quite hot during use.
- ✓ Take care to avoid broken glass and sharp objects at all times.
- ✓ Gloves will be provided when you enter the lab in order to prevent direct contact of the organic solvents with your skin. Wear goggles at <u>all</u> times.
- ✓ There are flammable and volatile liquids in use.

DISPOSAL

- ✓ Wash excess aqueous solutions down the drain with copious amounts of water.
- ✓ Organic solvents are to be placed in the organic liquid waste container.
- ✓ Place excess organic solids in the solid waste container.
- ✓ Dispose of soiled gloves and paper towels in the proper waste container.
- ✓ Dispose of broken glass in the broken glass container.

CHEM 332L Experiment 7: Claisen Condensation

Procedure

REMINDER: Close and cap all reagent and waste containers.

1. Thoroughly mix 1.6 g of KOtBu and 3.2 mL of ethyl phenylacetate in a 25 mL round bottom flask. Stir thoroughly so that you make a slurry.

2. Attach the reflux condenser to the 25 mL round bottom flask. Make sure the apparatus is secure.

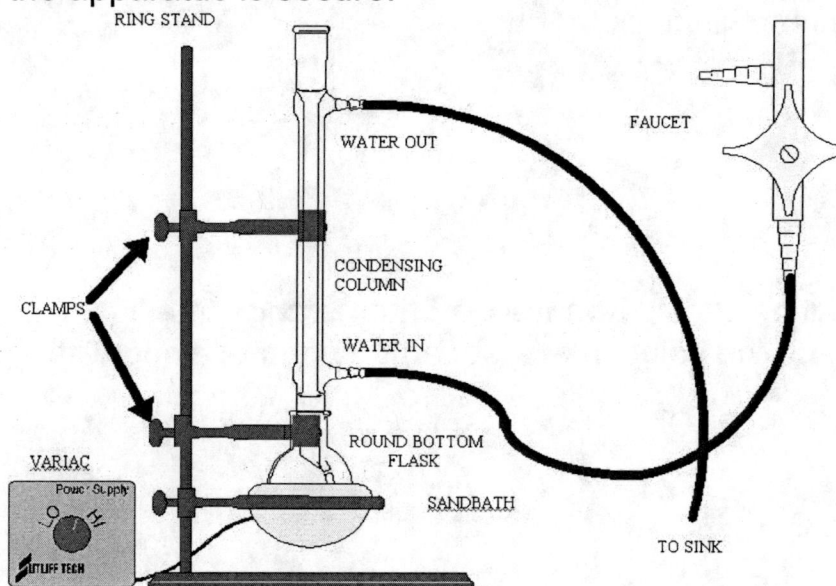

3. Heat the reaction mixture on a heating mantle and reflux for 30 minutes.

4. Neutralize the reaction mixture by **slowly** adding 3 mL of 3M HCl. It will not require all 3mL so make sure to check pH constantly to not make it acidic.

5. Test the solution periodically with pH paper to ensure complete neutralization. Perform this test by stirring the mixture with a glass rod and placing the rod on the pH paper. **DO NOT** stick the pH paper into the reaction flask.

6. The solution should turn from a bright orange to a light cloudy yellow color. Add the solution to a separatory funnel along with 7 mL of ether.

CHEM 332L Experiment 7: Claisen Condensation

7. Mix the two phases well and remove the aqueous layer and set it aside. **Be sure to vent frequently!** Collect the organic phase in an Erlenmeyer flask. Return the aqueous phase to the separatory funnel and extract with another 7 mL aliquot of ether. Combine the two organic phases into the Erlenmeyer flask. **There should be no visible water layer in your flask.**

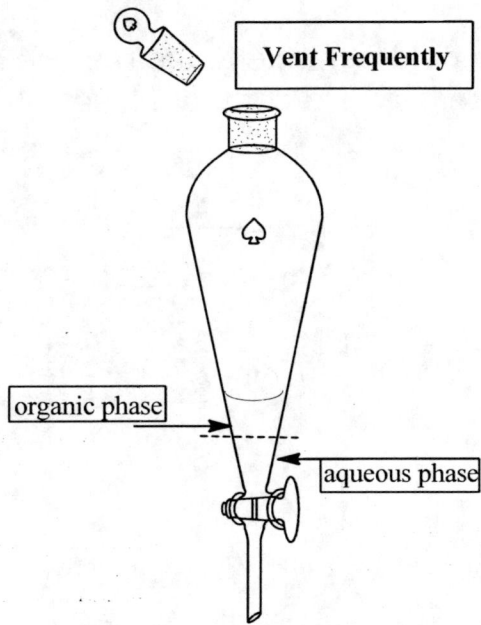

8. Dry the organic extract by adding a spatula of anhydrous MgSO$_4$ leaving the MgSO$_4$ hydrate in the container. Swirl and decant the ether mixture into a 50 mL Erlenmeyer flask.

$$\text{Et}_2\text{O} + \text{Product} + X \cdot H_2O \xrightarrow{\text{MgSO}_4} \text{Et}_2\text{O} + \text{Product} + \text{MgSO}_4 \cdot X \cdot H_2O$$

9. Evaporate the ether off by using a steam bath. You should continue to boil the ether off until a pale oily liquid remains.

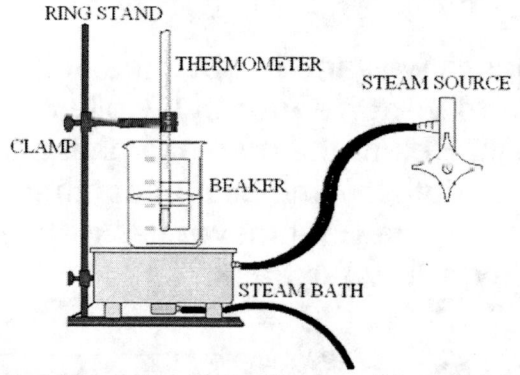

10. Add 5 mL of **ice cold** pentane to the residue and mix thoroughly. Place the mixture in an ice bath for at least 10 minutes. Filter off the solid product.

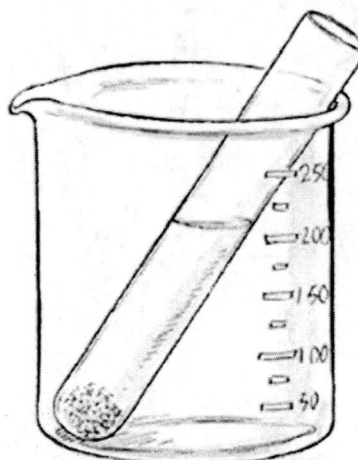

11. Weigh and record the mass of the crude product.

12. Recrystallize the product using hexane, filter by suction. Remove the product from the Buchner funnel and place on a new piece of filter paper. Cover with an inverted beaker, air dry, and calculate the percent recovery.

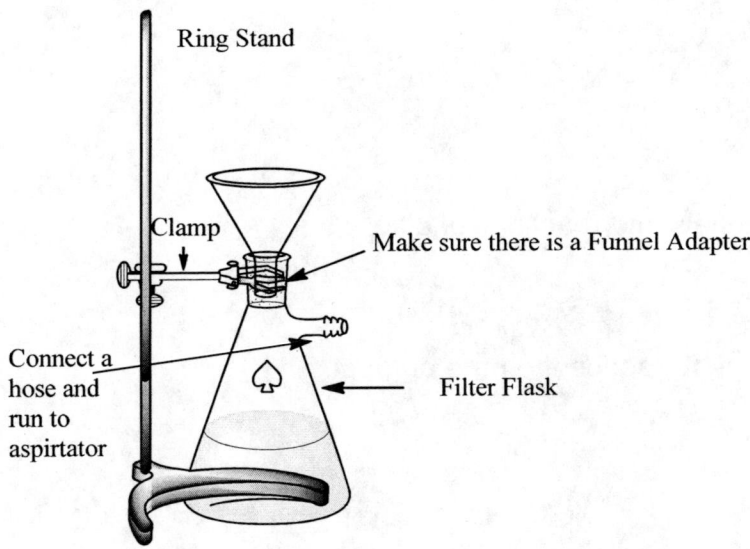

13. Weigh the product, obtain the melting point, and calculate percent yield of the pure product.

OBSERVATIONS

1. Describe the starting materials and reaction conditions.

2. Describe the crude product. Describe the pure product.

Own Observations:

CHEM 332L Experiment 7: Claisen Condensation

RESULTS

Mass of ethyl phenylacetate = _____ g

Theoretical Yield of Product = _____ g

Measured Mass of Product = _____ g

Percent Yield = _____ %

Percent Recovery = _____ %

Melting Point = _____ °C

Theoretical Melting Point _____ 75-78 °C

Percent Difference = _____ %

CALCULATIONS

Conclusions

Experiment Eight 8

EAS: Nitration of Methyl Benzoate

OBJECTIVES

1. Microscale synthesis of meta-nitromethylbenzoate.
2. Isolate and recrystallize crude product from methanol.
3. Calculate the theoretical yield.
4. Record mass and calculate percent yield of crude product.
5. Record mass, melting point and calculate percent yield and percent recovery of the pure product.

THEORY

1. EAS: Why Subsitution vs. Addition

    a. A typical Alkene undergoes a straight addition reaction. However, due to the high electron density within the benzene ring, this simple addition does not occur. What is actually seen is a substitution reaction.

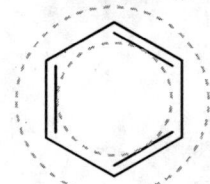

6 π electrons; high electron density.

An aromatic system has a <u>high electron density</u> and therefore attracts electrophiles (E⁺).

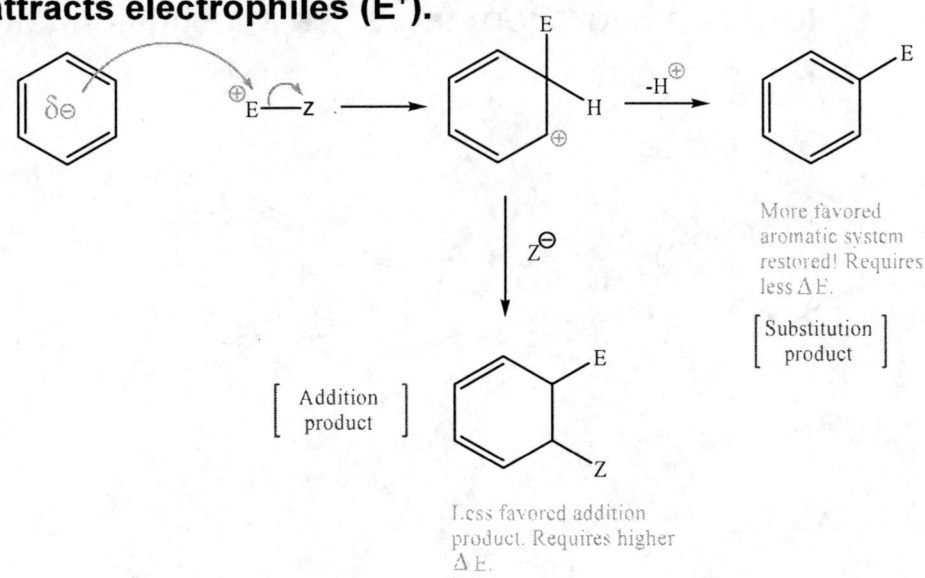

More favored aromatic system restored! Requires less ΔE.

[Substitution product]

[Addition product]

Less favored addition product. Requires higher ΔE.

287

2. Mechanism of an EAS reaction

STAGE 1 Formation of the electrophile.

$$E\text{---}Z \longrightarrow E^{\oplus} + Z^{\ominus}$$

STAGE 2 Addition of the electrophile to the aromatic ring.

High energy intermediate; Loss of Resonance energy.

STAGE 3 Restoring aromatic stability and resonance energy – loss of a hydrogen, therefore substitution takes place.

CHEM 332L Experiment 8: EAS: Nitration of Methyl Benzoate

3. **Activators and Deactivators:** Certain groups on the benzene can result in an activating or deactivating affect on the benzene ring. Groups that have a lone unshared pair of electrons will be able to donate these electrons to the resonance of the benzene ring. This in turn causes the ring to become activating.

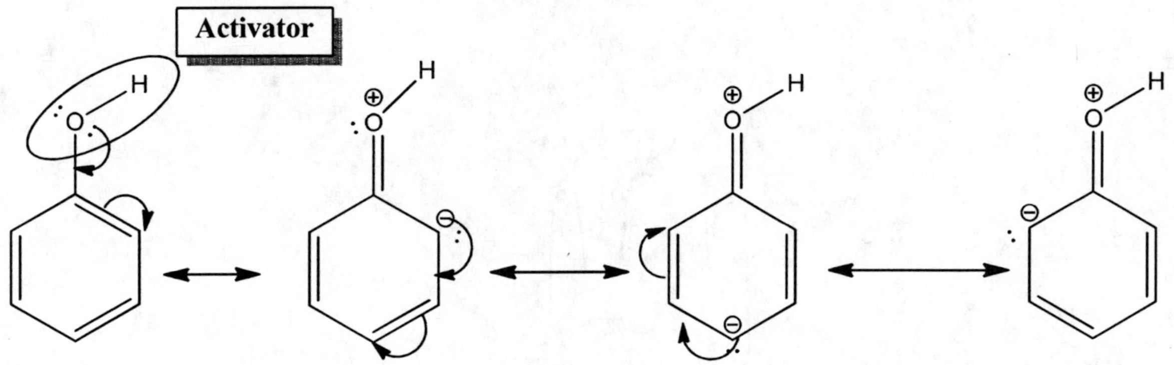

The ester group is an electron-withdrawing group as shown below. Phenols are more reactive than methyl benzoate because of the electron-donating group (EDG). Carbonyl groups are electron withdrawing and place a positive charge on the ring, decreasing the rate of reaction when compared to benzene.

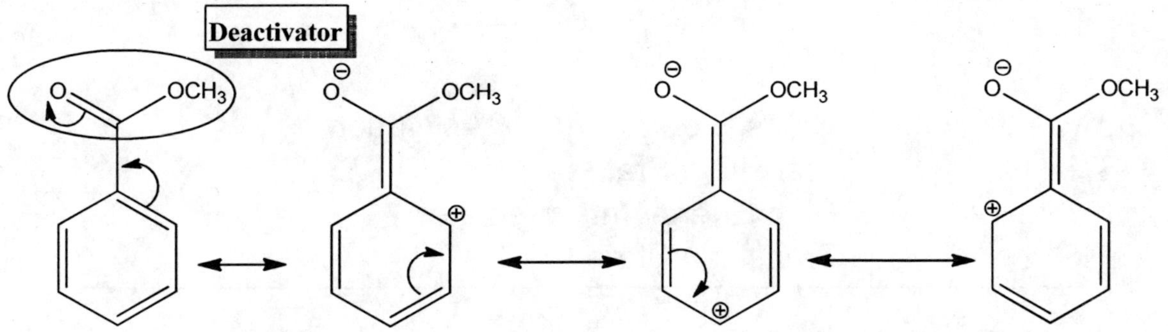

CHEM 332L Experiment 8: EAS: Nitration of Methyl Benzoate

A General Mechanism for an EAS Reaction

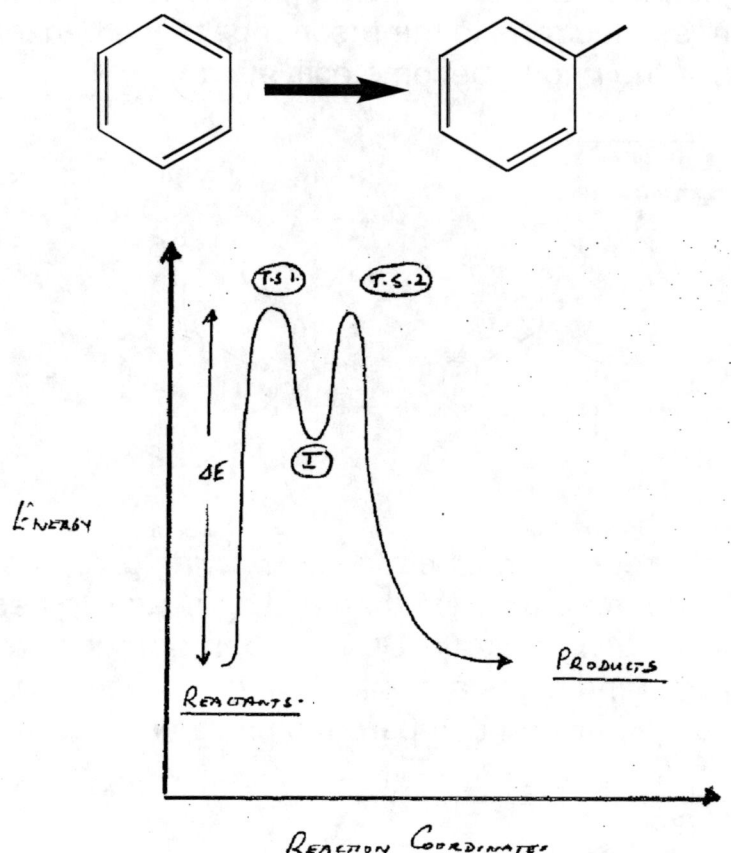

Δ E	≡	Gibbs Free Energy of Activation
T.S. 1	≡	Transition State I
I	≡	High Energy Intermediate
T.S. 2	≡	Transition State II

Mechanism for the Nitration of Methyl Benzoate

2. Formation of the Nitronium Ion

$$HNO_3 + H_2SO_4 \longrightarrow {}^{\oplus}NO_2 + H_2O + HSO_4^{\ominus}$$

$$1 \quad : \quad 3$$

A combination of Nitric acid [concentrated] and concentrated sulfuric acid in a ratio of 1:3 is used to generate the <u>nitronium ion</u>. This mixture is called the <u>nitrating mixture</u> – an extremely corrosive combination of liquids.

Step 1: Formation of Electrophiles

Step 2: Addition to the Aromatic System, Rate Determining Step

Step 3: Restoration of Aromaticity, Fast Step

CHEM 332L Experiment 8: EAS: Nitration of Methyl Benzoate

Mechanism of EAS
a. Slow and fast steps (discuss with energy diagram)
b. Transition states and resonance structures

Breaking aromaticity

Restoring aromatic stability

Mechanistic Explanation why EAS takes place at the META position.

EXAMPLE:

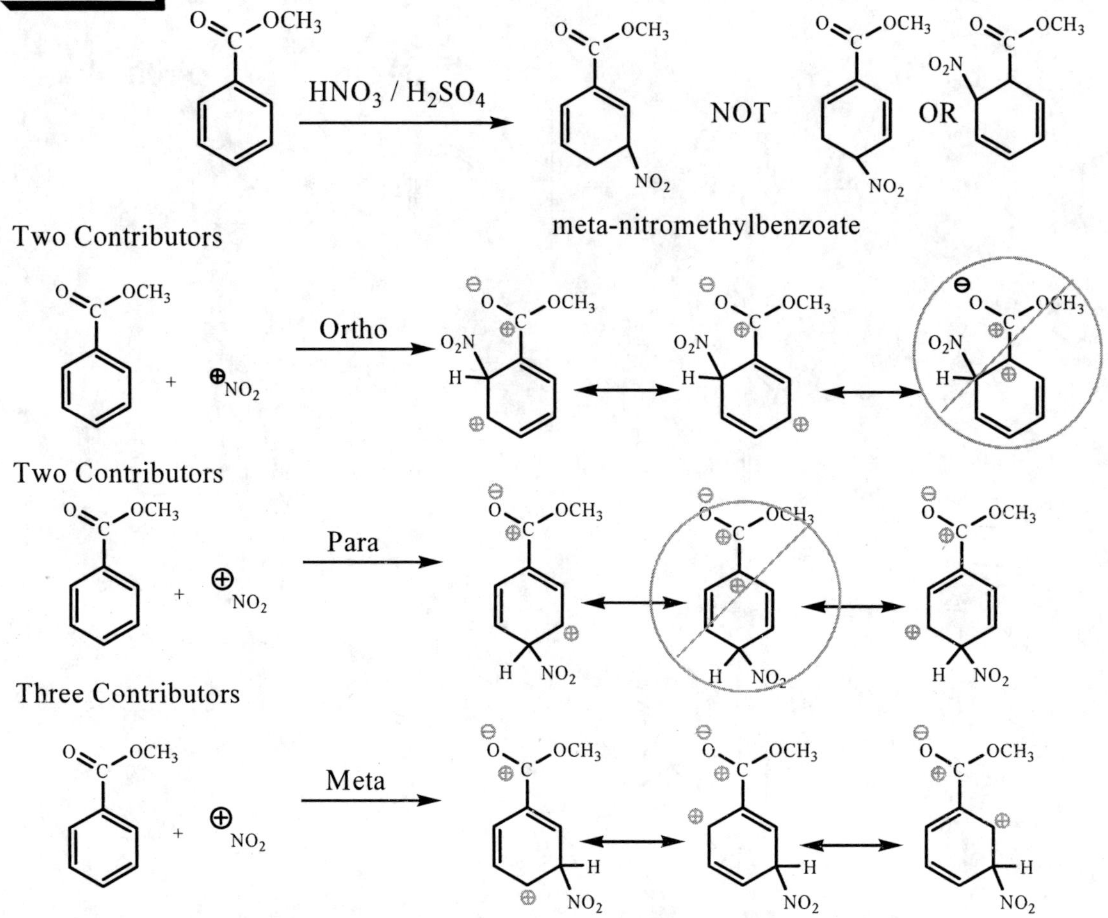

Methyl benzoate + HNO$_3$ / H$_2$SO$_4$ → meta-nitromethylbenzoate (NOT para OR ortho)

Two Contributors — Ortho attack: three resonance structures shown, with the one bearing positive charge on the ipso carbon (adjacent to the electron-withdrawing ester group) crossed out as a poor contributor.

Two Contributors — Para attack: three resonance structures shown, with the middle one (positive charge on the ipso carbon) crossed out as a poor contributor.

Three Contributors — Meta attack: all three resonance structures are major contributors (none place the positive charge on the ipso carbon bearing the ester group).

Inference:

Ortho / para substitution provides only <u>two major contributors</u> structures, which leads to a higher ΔG_0 for that pathway.

On the contrary, meta substitution provides <u>three major contributors,</u> which significantly <u>lowers</u> ΔG_0. Therefore meta substitution is favored.

CHEM 332L Experiment 8: EAS: Nitration of Methyl Benzoate

REAGENTS

Name, Structure, MW (g/mol)	Melting °C	Boiling °C	Density g/mL	Properties	Safety
Methyl Benzoate $C_8H_8O_2$ 136.15	-12.5	199.6	1.088	Colorless, oily liquid	Irritating to the eyes, mucous membranes and upper respiratory tract; when heated to decomposition it emits toxic fumes of CO and CO_2
Nitric acid HNO_3 63.012	-42	120.5	1.413	Colorless, yellow, or red fuming liquid with an acrid, suffocating odor; Corrosive material	Highly corrosive to the eyes, skin, and mucous membranes; powerful oxidizing agent that ignites on contact or reacts explosively with many organic and inorganic substances
Methanol CH_3OH 32.04	-98 °C	64.6 °C	0.791	Clorless liquid with a characteristic pungent odor	May cause severe skin and eye irritation; it can be absorbed through the skin; may cause narcosis; ingestion may cause blindness
Sulfuric Acid H_2SO_4 98.07	3	290	1.84	Corrosive material; Colorless (pure) to dark brown, oily, dense liquid with sharp, acrid odor; HYGROSCOPIC	Highly corrosive; causes severe, deep burns on eye and skin contact and upon inhalation of sulfuric acid mist; highly reactive - reacts violently with many organic and inorganic substances.
3-Nitromethyl Benzoate $C_8H_7NO_4$ 181.14	78-80	279	-	-	-

SAFETY

- ✓ As a <u>steambath</u> is in use, extra precaution must be taken.
- ✓ The apparatus can also get quite hot during use.
- ✓ Take care to avoid broken glass and sharp objects at all times.
- ✓ Gloves will be provided when you enter the lab in order to prevent direct contact of the organic solvents with your skin. Wear goggles at <u>all</u> times.
- ✓ Nitric acid is extremely corrosive.
- ✓ There are flammable and volatile liquids in use.

DISPOSAL

- ✓ Wash excess aqueous solutions down the drain with excess water.
- ✓ Place organic solvents in predesignated organic liquid waste container separate from other organic liquid waste.
- ✓ Dispose of soiled gloves and paper towels in trash. Dispose of broken glass in the *broken glass container*.
- ✓ Dispose of the product in the solid organic waste container.

PROCEDURE

REMINDER: Close and cap all reagent and waste containers.

1. Place 0.6 mL of sulfuric acid in a 25 mL Erlenmeyer flask.

2. Carefully add 0.3 mL of <u>methyl benzoate</u> and place the mixture in an ice bath.

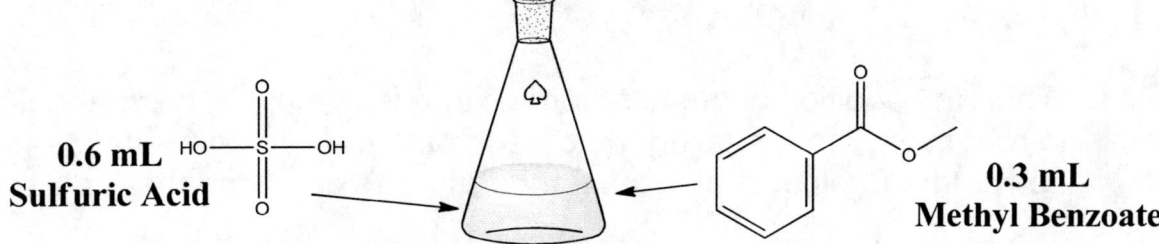

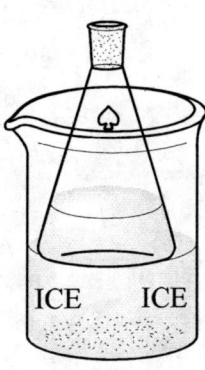

3. Very carefully add a cold mixture of 0.2 mL of <u>nitric acid</u> using a micropipette. Swirl the mixture during the addition being sure not to leave the mixture out of the ice bath for extended periods.

$$HNO_3 + H_2SO_4 \longrightarrow {}^{+}NO_2 + H_2O + HSO_4^{-}$$

$$1 \quad : \quad 3$$

4. When the addition of the nitration mixture is complete, warm the flask to room temperature using the steam bath at a very low setting. Leave for 20 minutes at room temperature to complete the reaction.

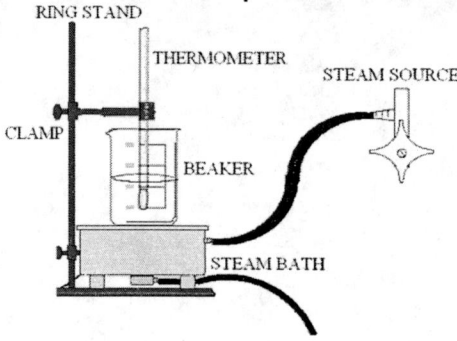

5. Add 1 to 2 small ice cubes to the mixture in the test tube. A solid product should precipitate at this point. Leave for 10 minutes to complete the precipitation process.

6. Collect the product using suction filtration. Rinse product with a few drops of ice cold water. Rinse again with the dropwise addition of two 1 mL portions of ice cold methanol. Drain thoroughly.

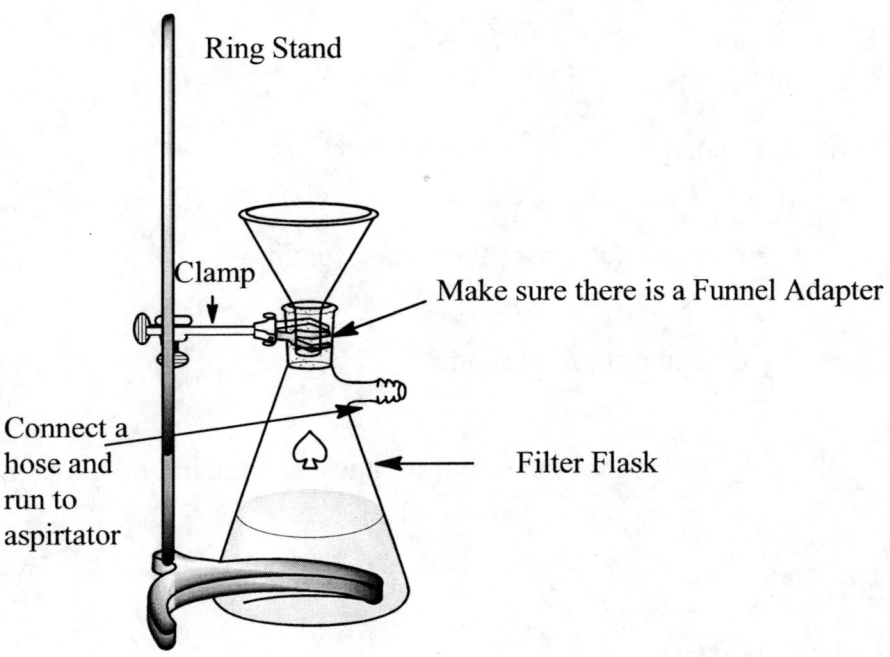

7. Transfer the product to a pre-weighed container. Record the mass and the melting point. Calculate the percent yield.

8. Recrystallize the product from methanol.
   a. Transfer the crude product to a clean test tube.
   b. Cover the solid with a minimal volume of methanol.
   c. Heat gently on the steam bath and agitate frequently.
   d. Add methanol dropwise until most of the solid has dissolved.
      Note: The product is slightly soluble in methanol, so use methanol very sparingly.
   e. Remove the test tube from the steam bath and cool at room temperature for 10 minutes then place it in an ice bath to complete the crystallization.
   f. Record the mass of the pure product, calculate percent yield, and determine the melting point.

CHEM 332L Experiment 8: EAS: Nitration of Methyl Benzoate

OBSERVATIONS

1. Describe the methyl benzoate.

2. Describe how the boiling changes over the course of time.

3. Describe any odors you may detect from afar.

4. Describe any color changes in the round bottom flask as reaction progresses.

5. Describe the product.

Own Observations:

_____
_____
_____
_____
_____
_____
_____
_____
_____
_____
_____
_____
_____
_____
_____
_____

RESULTS

Mass of methyl benzoate = _____ g

Theoretical Yield = _____ g
(nitro-methyl benzoate)

Measured Mass of Crude Product = _____ g
(nitro-methyl benzoate)

Percent Yield = _____ %

Mass of Recovered Product = _____ g
(nitro-methyl benzoate)

Percent Recovery = _____ %

Melting Point Pure Product = _____ °C

Percent Difference = _____ %

CALCULATIONS

Conclusions

Experiment Nine

Electrophilic Aromatic Synthesis: Friedel-Crafts Alkylation

OBJECTIVES
1. Synthesize 1,4-di-*t*-butyl-2,5-dimethoxybenzene.
2. Isolate the crude product and recrystallize from methanol.
3. Calculate the theoretical yield.
4. Record the mass, obtain a melting point, and calculate the percent yield.

THEORY
1. General
   a. Lewis acidity and basicity
   A Lewis acid is a species that accepts a pair of electrons to form a new covalent bond. A Lewis base is a species that donates a pair of electrons to form a new covalent bond. The electron pair becomes shared with another atom to form a covalent bond.

   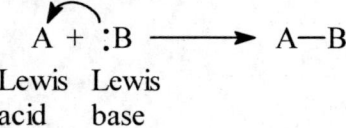

   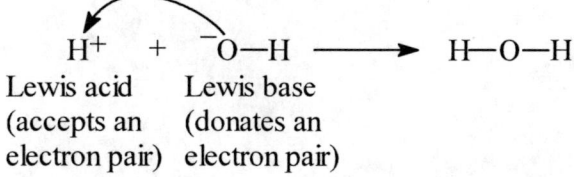

   b. Catalyst
   A catalyst is a substance that increases the rate of a reaction, but is not consumed in the reaction (it is still present after the product is formed).
   Sulfuric acid is the Lewis acid catalyst for this experiment.

c. Carbocations and their stability

$$\text{Stability: } 3°>2°>1° \text{ of } C^+$$

The larger the area the positive charge is spread over, the greater stability of the cation. In this experiment, the carbocation produced is trimethyl carbocation. This is a very stable carbocation.

CHEM 332L Ex 9: Electrophilic Aromatic Synthesis: Friedel-Crafts Alkylation

2. Reaction
a. Overall reaction

1,4-Dimethoxybenzene + 2 $CH_3-C(CH_3)_2-OH$ (t-Butyl alcohol) $\xrightarrow{H_2SO_4 \text{ or } H_3PO_4}$ 1,4-Di-t-butyl-2,5-dimethoxybenzene

b. Reaction mechanism

STEP 1   Formation of the carbocations (electrophiles).

STEP 2   Addition of the electrophile to the reactive benzene ring at the ortho position with respect to one of the alkoxy groups, giving the high energy intermediate.

STEP 3   Loss of a hydrogen atom to reform the more stable π electron aromatic system.

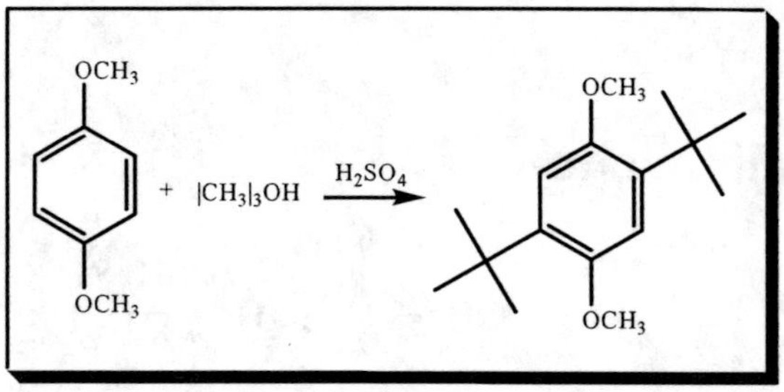

CHEM 332L Ex 9: Electrophilic Aromatic Synthesis: Friedel-Crafts Alkylation

$$\text{1,4-dimethoxybenzene} + (CH_3)_3OH \xrightarrow{H_2SO_4} \text{2,5-dimethoxy-di-tert-butylbenzene}$$

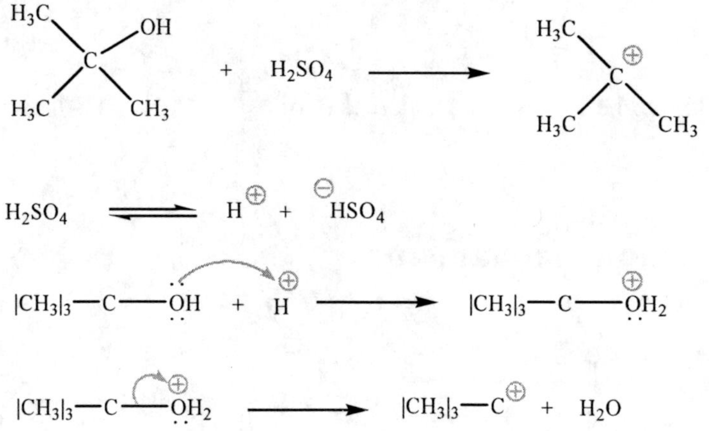

A

$(CH_3)_3C-OH + H_2SO_4 \longrightarrow (CH_3)_3C^{\oplus}$

$H_2SO_4 \rightleftharpoons H^{\oplus} + {}^{\ominus}HSO_4$

$(CH_3)_3-C-\ddot{O}H + H^{\oplus} \longrightarrow (CH_3)_3-C-\overset{\oplus}{O}H_2$

$(CH_3)_3-C-\overset{\oplus}{O}H_2 \longrightarrow (CH_3)_3-C^{\oplus} + H_2O$

B

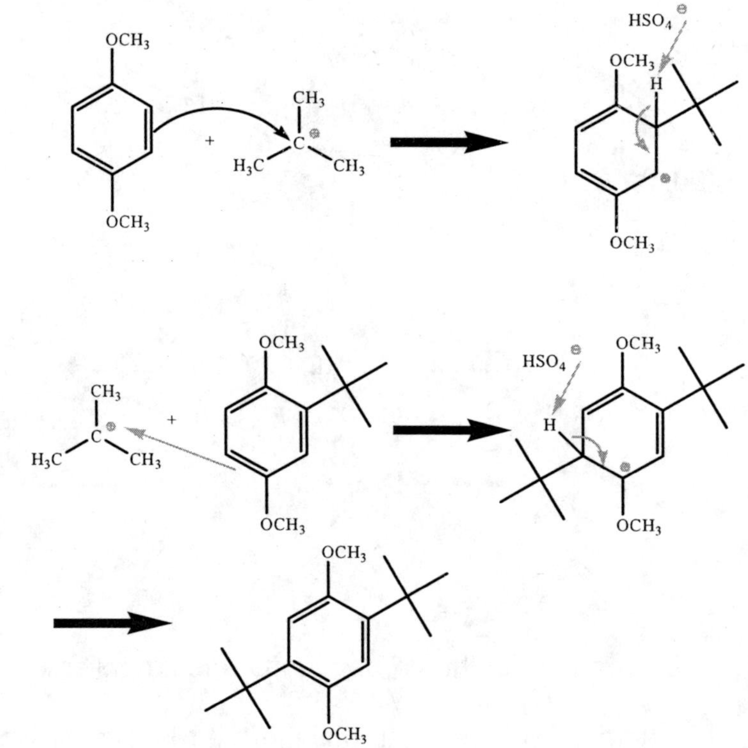

[Mechanism: 1,4-dimethoxybenzene + tert-butyl cation → arenium intermediate → HSO₄⁻ deprotonation → mono-tert-butyl-1,4-dimethoxybenzene → second tert-butyl cation attack → second arenium intermediate → final product 2,5-di-tert-butyl-1,4-dimethoxybenzene]

306

CHEM 332L Ex 9: Electrophilic Aromatic Synthesis: Friedel-Crafts Alkylation

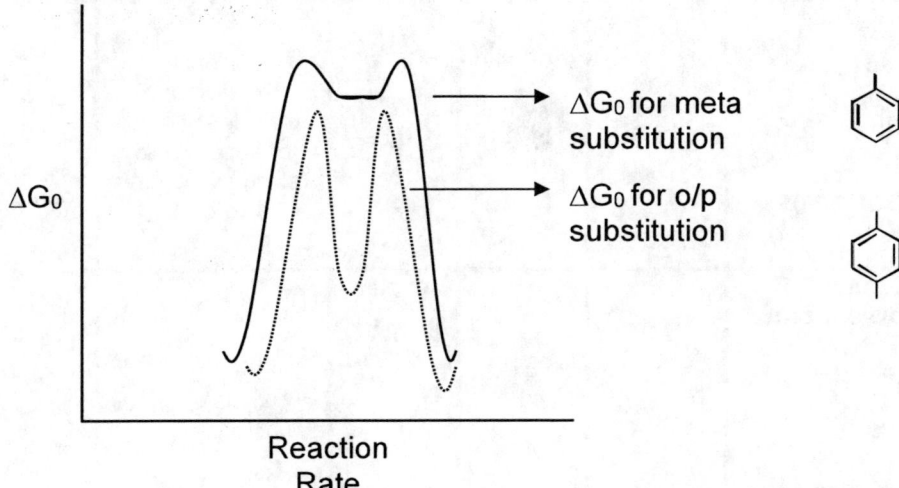

Ortho / Para substitution gives four major contributors which help stabilize the high energy intermediate, thus lowering the energy of activation for the slow rate determining step.

CHEM 332L Ex 9: Electrophilic Aromatic Synthesis: Friedel-Crafts Alkylation

REAGENTS

Name, Structure MW (g/mol)	Melting °C	Boiling °C	Density g/mL	Properties	Safety
1,4-Dimethoxybenzene $C_8H_{10}O_2$ 138.16	56 - 60	212.6	1.053	Slightly soluble in water; White crystals	IMMEDIATELY flood affected skin with water while removing and isolating all contaminated clothing; gently wash all affected skin areas thoroughly with soap and water
t-Butanol C_4H_9OH 74.12	25.5	82.2	0.786	Flammable Liquid; Colorless liquid with a camphor-like odor; solid in cold weather	This compound is highly toxic by inhalation; irritating to the skin, eyes and mucous membranes; can be narcotic in high concentrations
Methanol CH_3OH 32.04	-98 °C	64.6 °C	0.791	Colorless liquid with a characteristic pungent odor	May cause severe skin and eye irritation; it can be absorbed through the skin; may cause narcosis; ingestion may cause blindness
Sulfuric acid H_2SO_4 98.073	3	290	1.84	Corrosive material; Colorless (pure) to dark brown, oily, dense liquid with sharp, acrid odor; HYGROSCOPIC	Highly corrosive; causes severe, deep burns on eye and skin contact and upon inhalation of sulfuric acid mist; highly reactive, reacts violently with many organic and inorganic substances
Acetic acid CH_3COOH 60.05	16.6	117.9	1.0492	Corrosive material; Colorless liquid or solid with a strong vinegar-like odor	Corrosive to tissue; irritating to the skin, mucous membranes, upper respiratory tract and eyes; fatal if swallowed; causes severe burns to all tissues contacted; when heated to decomposition, it emits toxic fumes of CO and CO_2
1,4-di-*t*-butyl-2,5-dimethoxybenzene $C_{16}H_{26}O_2$ 250.38	104 - 105	-	-	White solid	Eye irritant

SAFETY
- As a steambath is in use, extra precaution must be taken.
- Take care to avoid broken glass and sharp objects at all times.
- As with all acids and bases, extra precaution should be taken when sulfuric acid is in use.
- Glacial acetic acid is very concentrated. Do not inhale and avoid skin contact.
- Gloves will be provided when you enter the lab in order to prevent direct contact of the organic solvents with your skin. Wear goggles at <u>all</u> times.

DISPOSAL
- Wash excess aqueous solutions down the drain with excess water.
- Organic solvents are to be placed in predesignated organic liquid waste container separate from other organic liquid waste (t-butanol, methanol).
- Dispose of soiled gloves and paper towels in trash. Dispose of broken glass in the *broken glass container*.
- Dispose of the product and excess 1,4-dimethoxybenzene in the solid organic waste container.

CHEM 332L Ex 9:Electrophilic Aromatic Synthesis: Friedel-Crafts Alkylation

PROCEDURE

REMINDER: Close and cap all reagent and waste containers.

1. Dissolve 120 mg of 1,4-dimethoxybenzene in 0.4 mL of **glacial acetic acid** in a small test tube. Warm if necessary.

2. Add 0.2 mL of t-butanol to the test tube.

3. Cool in an ice water bath.

4. Add 0.6 mL of concentrated sulfuric acid dropwise down the side of the test tube, stir the solution thoroughly with a glass rod after each drop added.

5. Plug the test tube with a test tube stopper.

6. Place a beaker of water on the steam bath and maintain a water temperature of around 35 -40° C. Place the test tube in the warm water bath.

7. Check the temperature of the bath occasionally with the thermometer and maintain the desired temperature for 30 - 40 minutes.

8. Let the test tube cool for 5 minutes and place the mixture in an ice bath to induce crystallization.

9. *Cautiously* add 1 drop of water to the reaction mixture and stir. Continue adding water until the total volume in the tube is about 3.5 mL.

CHEM 332L Ex 9:Electrophilic Aromatic Synthesis: Friedel-Crafts Alkylation

10. Remove and discard the supernatant liquid from the cold reaction mixture using a Pasteur pipette. *Caution*: do not disturb the solid.

11. Wash solid product with 2 mL of *ice cold* water. Remove and discard the water with a Pasteur pipette.

12. Wash the solid with a 0.2 mL portion of *ice cold* methanol.

13. The product is slightly soluble in methanol; use methanol very sparingly. Recrystallize the product as follows:
    a. Cover the solid with a minimal amount of methanol
    b. Heat gently on the steam bath and agitate frequently.
    c. Add methanol dropwise until most of the solid has dissolved.
    d. Remove the test tube from the steam bath and cool at room temperature for 10 minutes, then place it in an ice bath to complete the crystallization.

14. Collect product by suction filtration. Air dry the product using suction filtration for 5 minutes.

15. Record the mass, calculate the percent yield, and determine the melting point.

APPARATUS

STEAM BATH

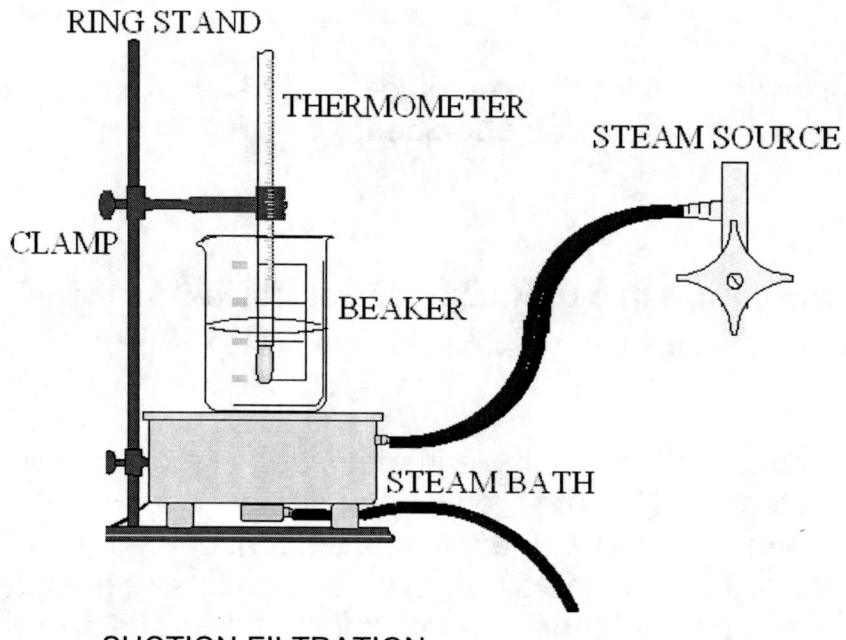

SUCTION FILTRATION

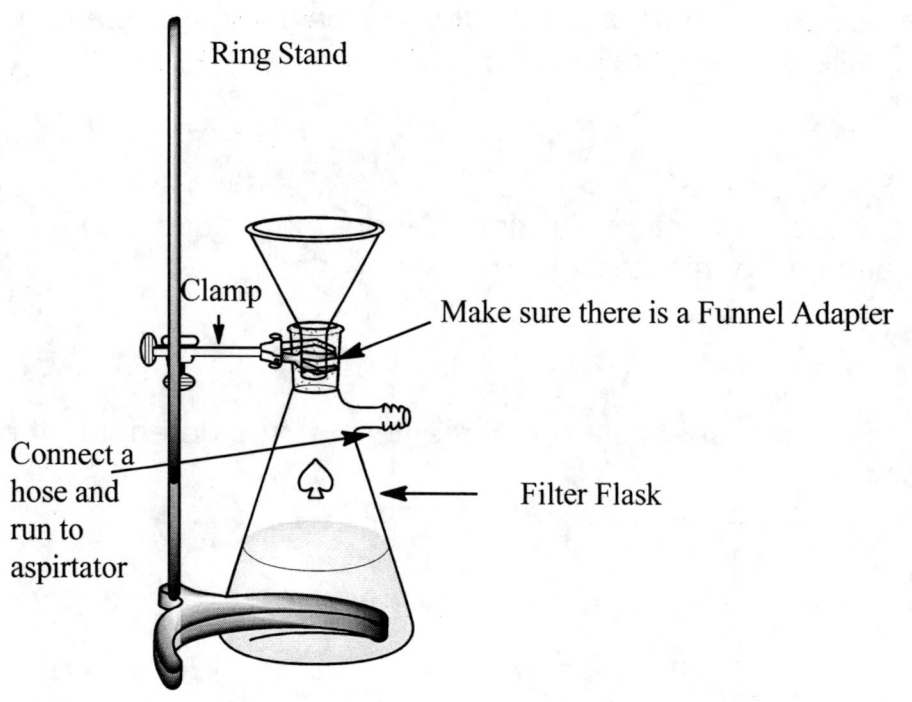

CHEM 332L Ex 9:Electrophilic Aromatic Synthesis: Friedel-Crafts Alkylation

OBSERVATIONS

1. Describe the starting material 1,4-dimethoxybenzene.

2. Describe how the boiling changes over the course of time.

3. Describe any odors you may detect from afar.

4. Describe any color changes in the round bottom flask as reaction progresses.

5. Describe the product.

Own Observations:

_____
_____
_____
_____
_____
_____
_____
_____
_____
_____
_____
_____
_____
_____

RESULTS

Mass of 1,4-dimethoxybenzene = _____ g

Theoretical Yield = _____ g
(1,4-di-*t*-butyl-2,5-dimethoxybenzene)

Measured Mass of Product = _____ g
(1,4-di-*t*-butyl-2,5-dimethoxybenzene)

Percent Yield = _____ %

Melting Point Product = _____ °C

CALCULATIONS

Conclusions

Experiment Ten

Click Reaction

OBJECTIVES

1. Utilize "Click" chemistry to synthesize 1,2,3 Triazoles from Azides and Terminal Acetylenes.
2. Monitor reaction synthesis via TLC and calculate percent yield.
3. Perform IR and NMR analysis of the product.

THEORY

The term "Click chemistry" refers to a group of simplified chemical reactions that signify the joining of two molecular pieces no matter what other attachments may be on each component. The methodology was recognized by Nobel Laureate Professor Barry K. Sharpless and his group at the Scripps Research Institute. Some properties of "Click" reactions are as follows:

- easy to perform
- intended products are found in very high yields with little or no by-products
- work well under many conditions (usually especially well in water)
- are unaffected by the nature of the groups being connected to each other

Certain reactions exemplify the philosophy behind "Click" chemistry. For example, the copper catalyzed Azide-Alkyne cycloaddition reaction which was introduced by Sharpless and Meldal in 2002 is considered by most scientists as the "click" reaction. A schematic of the overall reaction is shown below:

Copper Catalyzed Azide-Alkyne Cycloaddtion Reaction

Image adapted from: http://clickchemicals.com/CuAAC.html

The utilization of the term "click" is simply a means to explain how easy these reactions demonstrate the joining of two specific molecular pieces. However, the utilization of copper as a catalyst may not always be the preferred choice by scientists. The reactions shown below have been identified as using "click" chemistry without the addition of copper.

Uncatalyzed Cycloaddtion Reaction

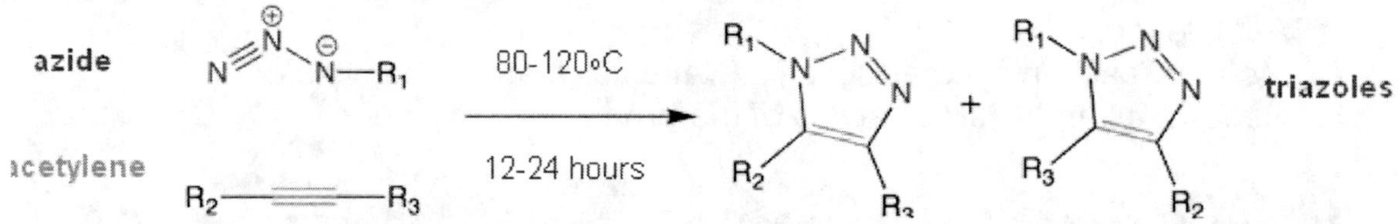

azide: $N \equiv \overset{\oplus}{N} - \overset{\ominus}{N} - R_1$

acetylene: $R_2 - \equiv - R_3$

80-120 °C, 12-24 hours →

triazoles (1,4 and 1,5 isomers with R_1 on N, R_2 and R_3 on carbons)

Image adapted from: http://clickchemicals.com/Uncatalyzed_Cycloaddition.html

Note: Thermodynamically favored reaction, however multiple products are formed and high temperatures are required.

Strain Promoted Azide-Alkyne Cycloaddtion Reactions

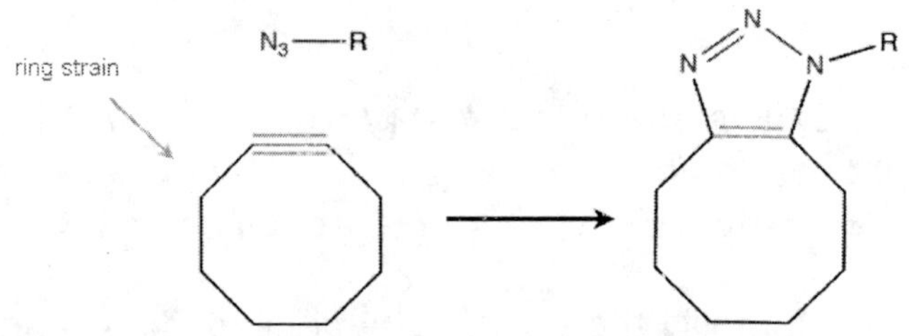

ring strain + N_3-R → fused triazole-cyclooctane product

Image adapted from http://clickchemicals.com/Strain_Promoted.html

Note: Reaction favors release of ring strain to form "click" product, however the reaction is slow. Rate is increased by adding electron withdrawing groups adjacent to the alkyne.

Other catalysts have been applied to these reactions as well. Scientists have utilized ruthenium complexes as a "click" catalyst and have found that 1,5 triazoles are formed as a result. The reaction is shown below:

CHEM 332L Experiment 10: Click Reaction

Ruthenium Complex Catalyzed Azide-Alkyne Cycloaddition Reactions

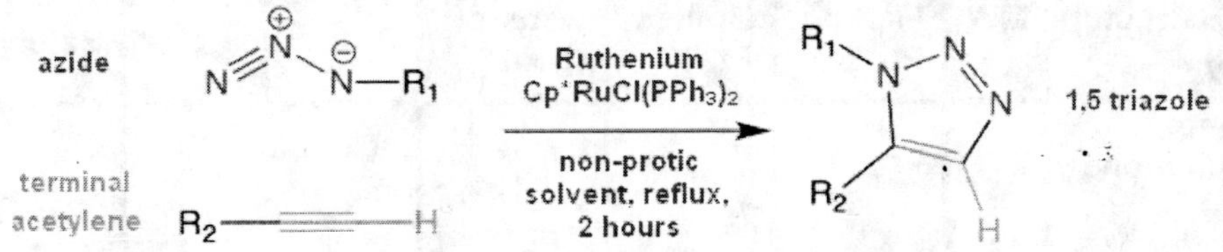

azide + terminal acetylene R_2—≡—H → (Ruthenium Cp*RuCl(PPh$_3$)$_2$, non-protic solvent, reflux, 2 hours) → 1,5 triazole

Image adapted from: http://clickchemicals.com/Ruthenium_Catalyzed.html

Note: Internal alkynes can also be utilized in this reaction. However, high temperature and the use of an aprotic solvent (i.e. benzene) are necessary.

The experiment for today will utilize benzyl azide and various acetylenes. Copper (I) and Sodium Ascorbate will be added to catalyze the "click" reaction. The overall reaction and mechanism is shown below:

General Click Reaction Mechanism:

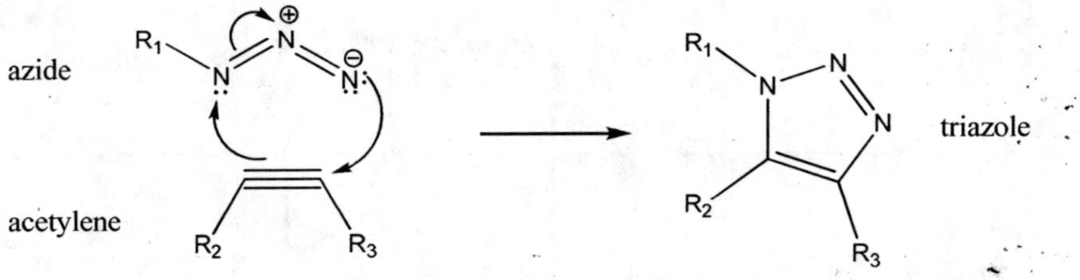

azide + acetylene (R_2, R_3) → triazole (R_1, R_2, R_3)

REAGENTS

Name, Structure, MW (g/mol)	Melting Pt. °C	Boiling Pt. °C	Density (g/mL)	Properties	Safety
2-Bromo-acetophenone C_8H_7BrO 199.05	46-51	135	-	Very low water solubility	Corrosive lachrymator; danger to the eyes and mucous membranes
Phenyl Propargyl Ether C_9H_8O 132.1592	-	-	1.030	Clear yellow liquid	Light and air sensitive
Copper Sulfate Pentahydrate $CuSO_4 \cdot 5H_2O$ 249.68	150	-	2.284	Blue crystalline solid	Irritant
Sodium Azide NaN_3 65.0099	275	-	1.846	White solid	Highly toxic, Explosive (make sure not to touch with anything Metal)
Sodium Ascorbate $C_6H_7CO_6Na$ 198.12	218	-	-	White or nearly white, crystalline powder	-
***tert*-butanol** $C_4H_{10}O$ 74.1216	25.69	82.4	0.78086	Colorless liquid; semi-solid	Skin irritant; harmful if inhaled

SAFETY

- ✓ **DO NOT** add boiling chips to a hot liquid. A superheated liquid may boil over or even explode.
- ✓ Take care to avoid broken glass and sharp objects at all times.
- ✓ Gloves will be provided when you enter the lab in order to prevent direct contact of the organic solvents with your skin. Wear goggles at <u>all</u> times.
- ✓ There are flammable and volatile liquids in use.

DISPOSAL

- ✓ Wash excess aqueous solutions down the drain with copious amounts of water.
- ✓ Organic solvents are to be placed in the organic liquid waste container.
- ✓ Place excess organic solids in the solid waste container.
- ✓ Dispose of soiled gloves and paper towels in the proper waste container.
- ✓ Dispose of broken glass in the broken glass container.

PROCEDURE

REMINDER: Close and cap all reagent and waste containers.

1. Obtain 6 mL of *tert*-butanol/water mixture (1:1) and place it in a 20 mL scintillation vial with a stir bar.
2. Carefully weigh and add to the scintillation vial 400 mg of 2-Bromoacetophenone.
3. Carefully weigh and add 140 mg of sodium azide to the vial.
4. Very gently swirl the vial to disperse the reagents in the solvent.
5. Weigh and add 40 mg of sodium ascorbate to the vial.
6. Obtain 100 µL of 1M aqueous copper (II) sulfate pentahydrate from your TA and add it to the reaction.
7. As soon as all of the reagents have been added to the vial, *immediately* spot the reaction on a TLC plate alongside the starting materials.
8. Repeat this process every 30 minutes to monitor the reaction progress.
9. If your reaction proceeds slowly, your reaction can be stored in the hood overnight. Consult your TA.
10. Once the reaction has gone to completion according to TLC, pour the contents of the vial into 10-15 mL of ice water.
11. Obtain 5 mL of 10% aqueous ammonia and add it to the mixture.
12. Stir the reaction for another 5 minutes, ensuring that the reaction is as homogenous as possible.
13. Collect the solid precipitate with a Buchner funnel and allow it to dry **completely**.

CHEM 332L Experiment 10:Click Reaction

14. Weigh the product, obtain the melting point, and calculate percent yield of the pure product.

IR/NMR ANALYSIS

1. Prepare a KBr pellet with the help of your TA and perform an IR analysis of your product.
2. Prepare a ^{1}H-NMR tube by dissolving your product in deuterated chloroform.

APPARATUS

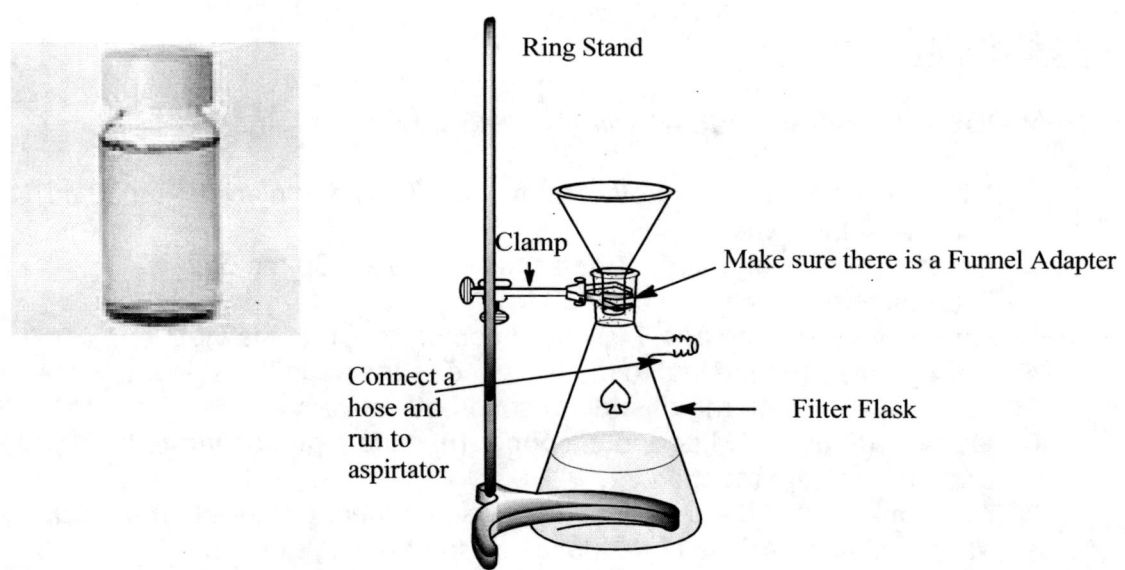

SCINTILLATION VIAL

SUCTION FILTRATION APPARATUS

OBSERVATIONS

Record observations that should be included in your report.

Own Observations:

RESULTS

Theoretical Yield of Product = _____ g

Measured Mass of Product = _____ g

Percent Yield = _____ %

Melting Point = _____ °C

Theoretical Melting Point _____ 149 °C

Percent Difference = _____ %

CALCULATIONS

CONCLUSIONS

Lab Checkout End of Semester

University of South Carolina Department of Chemistry & Biochemistry

CHEMISTRY 332L INVENTORY SHEET

Name: _____
TA: _____
Drawer #: _____

ITEM	QUANTITY IN	QUANTITY OUT
Round Bottom Flask (25 or 50 mL)	_____	_____
Side Arm Adapter (Distillation)	_____	_____
Thermometer	_____	_____
Thermometer Adapter	_____	_____
Condenser/Distilling Column	_____	_____
Graduated Cylinder (10 mL)	_____	_____
Graduated Cylinder (25 mL)	_____	_____
Beaker (50 mL)	_____	_____
Beaker (100 or 150 mL)	_____	_____
Beaker (250 mL)	_____	_____
Beaker (400 mL)	_____	_____
Separatory Funnel (250 mL)	_____	_____
Separatory Funnel Stopper	_____	_____
Buchner Funnel	_____	_____
Wide Mouth Short Stem Funnel	_____	_____
Spatula	_____	_____
Scoopula	_____	_____
Small Test Tubes (6)	_____	_____
Large Test Tubes (6)	_____	_____
Test Tube Rack	_____	_____
Test Tube Holder	_____	_____
Glass Stirring Rod	_____	_____
Erlenmeyer Flask (25 mL)	_____	_____
Erlenmeyer Flask (50 mL)	_____	_____
Erlenmeyer Flask (125 mL)	_____	_____
Erlenmeyer Flask (250 mL)	_____	_____
Filter Flask (125 or 250 mL)	_____	_____
Suction Filtration Adapter	_____	_____

Student Signature: _____

TA Signature: _____

APPENDIX SECTION

Appendix A – Techniques

Appendix B – General

Appendix C – Theoretical Yields

Appendix D – Spectroscopy

Appendix E – Grades Sheet for Labs

APPENDIX A – TECHNIQUES

CLEANING GLASSWARE

1. Glass assemblies are best cleaned **immediately** after each use.
2. Wipe off the grease on all ground glass joints with a Kimwipe or paper towel, soaked with acetone. This should be followed by washing with detergent and water.
3. A knowledge of the nature of the chemicals in your reaction flask helps a great deal in the proper clean-up of the glass apparatus. Water, soap and brush will do an excellent job on water-soluble or water-miscible compounds or substances. Organic material is easily removed by washing twice with a few milliliters of acetone. Tarry residues can be easily removed by soaking in acetone for 15 minutes, followed by a soap and hot water rinse.
4. The improper drying of your apparatus could affect the results of an experiment. A Grignard reaction, for example, will fail to start with trace amounts of water. For the routine drying of glassware, sprinkle a few milliliters of acetone after the regular wash, and then set aside on a drying rack. Towel dry glassware that has easily accessible surfaces, e.g., beakers, watch glasses, etc.
5. There is no need to dry glassware that will be used with aqueous solutions.

WASTE DISPOSAL

1. Organic waste should be collected in specially labeled containers.
2. Generally, aqueous solutions can be poured into the sink with ample water flushing.
3. Broken glassware, vials, and pipettes must be disposed of in a specially marked waste container.
4. Waste mercury, i.e., from broken thermometers, must be disposed of in a special container.

MAINTENANCE OF EQUIPMENT

1. The heating mantle should be wiped clean immediately after each use after cool. Do not immerse the apparatus in water.
2. The balances should be cleaned after each use. Any paper used in the process should be discarded.
3. Any chemical spills should be cleaned immediately.
4. At the end of a lab period, the lab bench should be cleaned with paper towels and washed with soap and water, if necessary. All waste should be discarded in the proper receptacles.

Appendix A

MEASURING MASS

Solids are conveniently massed over glassine paper (weighing paper) on the pan balance on the top loader. If the solid is an irritant or is sensitive to air or light, mass in a stoppered Erlenmeyer flask. Always remember to clean the balance area after each use and observe the following.

1. Do not weigh anything that is hot or warm.
2. Remember to tare the container prior to weighing.
3. Avoid drafts, bumps or excessive movements close to the balances. The volumes of liquids can be measured by means of a graduated cylinder, or more accurately with a volumetric flask or calibrated syringe. If the density of liquid is known, the weight can be calculated by using the simple relationship

$$\text{Mass (g)} = \text{density (g/mL)} \times \text{volume (mL)}$$

In reading volumes, always take the reading from the bottom of the meniscus at eye level. When dealing with small quantities of liquids, it is much more accurate to weigh them than to record a volume.

HANDLING CHEMICALS

Many chemicals are potentially quite dangerous. Observe the following guidelines when handling chemicals:

1. Use the hood for the transfer of volatile liquids and other dangerous chemicals.
2. Read the label carefully.
3. Never pipette anything by mouth.
4. Use rubber gloves when handling reactive reagents.
5. Hold the bottle away from your face.
6. Use a short stem funnel to prevent spillage.
7. Clean up spills, as instructed, immediately.
8. Never dip a contaminated pipette into a liquid container or an unclean spatula into a solid container.
9. Do not pour back any unused liquid or solid into reagent bottles. To avoid waste, pour out only what you need.
10. Most of the commonly used solvents are volatile, flammable and readily absorbed through the skin. Therefore:
    a) Use disposable gloves.
    b) Keep only a minimum quantity of flammable liquids on your bench.
    c) Keep solvents away from heat.
    d) Never heat solvents over a burner or hot-plate.
    e) Do not smell organic liquids.

Appendix A

11. Toxic solvents include aromatic hydrocarbons, organic acids, esters, halogenated hydrocarbons, carbon disulfide, methanol, and certain amines.
12. Rinse chemicals from flasks before overnight storage. If you must store chemicals in your locker, remember that besides contamination, disastrous consequences can occur, e.g., Tollens reagents have been known to explode upon overnight storage.
13. In general, if you have any doubt about the nature/toxicity hazard of any chemical you contemplate using, it is your responsibility to look up its properties. If you have any questions about any chemical, see your TA.

STORAGE OF CHEMICALS

Chemical compounds will deteriorate, decompose or form by-products if they are not purified and stored properly. In addition, improper storage may be a safety hazard. Observe the following guidelines for the proper storage of your compounds:

1. Store your compound immediately after completing the reaction work-up, preferably upon isolation of a pure product. Impure compounds or products that have not been worked-up properly tend to deteriorate quickly.
2. Leave products or reaction mixtures that need further work-up (and have no special storage recommendations) in your assigned area. Clearly label and stopper the container. Volatile liquids should be stored in special areas, i.e. your reaction product in ether should not be stored in your drawer. Consult your instructor if you have any questions.
3. If your compound is a liquid, store in a clean, tared and clearly labeled vial. Place a piece of aluminum foil underneath the cap (some liquids react with the inside of the cap) and store appropriately.
4. Dry solids may be stored in clean, dry, labeled vials and stoppered. Solids which are wet with solvent or water should be placed in a sample vial covered with foil which should be "perforated" in order that the solid will air dry. The vial should be placed in a small beaker. Under no circumstances should solids be placed on pieces of filter paper and left in the drawer.
5. Light-sensitive compounds should be stored in a dark bottle.
6. All vials that are submitted for inspection should be protected with "M" parafilm.
7. Never store any solvent, reagent or chemical in your drawer.

HEATING AND COOLING PROCEDURES

METHODS OF HEATING

A. **Steam Bath**
This is probably the safest method of heating. It is used for recrystallization or sublimation experiments and for warming organic liquid/solutions in

Erlenmeyer flasks or beakers. The utmost care should be used to avoid steam burns.

Please note the following:
1. The inlet hose (thick wall rubber tubing) should be firmly secured to the steam source and the inlet pipe on the steam bath by means of hose clamps.
2. The outlet pipe should be clamped on to an outlet hose which leads into the sink. For your own safety, make sure that these connections are correct before proceeding to turn on the steam source.

B. Hot-Plate/Stirrer
1. This is the most convenient method of heating since you have the added advantage of being able to stir your solutions magnetically as you heat. It is generally used for heating aqueous solutions/mixtures or for high-boiling organic liquids in Erlenmeyer flasks or beakers.

C. Heating Mantle
1. This device is designed for heating organic solutions/liquids in round bottom flasks, such as in the distillation of an organic liquid, or during a reaction (e.g., the Grignard reaction, the aldol condensation, etc.).
2. Heating Mantles come in different sizes to fit 50 mL, 100 mL, 250 mL, 500 mL round bottom flasks. The exact size and fit is important in maintaining efficient heat conduction.
3. The heating mantle is used in conjunction with a Variac voltage control. The following procedure should be followed when using a heating mantle.
    a. Use the right size heating mantle to fit the round bottom flask.
    b. Before you set up the apparatus, plug the heating mantle into the Variac and turn it to a setting of 30-40. Make sure that the Variac is plugged into the main power outlet. Cautiously touch the inside of the heating mantle to make sure that the unit/Variac is working. If the heating mantle does not warm up within 5-8 min, check the plugs to make sure the connections are tight. If this does not help, a fuse may have blown. Consult your TA.
    c. Never turn the Variac higher than 80 without consulting an instructor. The laboratory's electrical circuits may not handle the load if too many units are turned up high.

METHODS OF COOLING

Many reactions are too violent at elevated temperatures and thus are conducted at room temperature and occasionally at substantially lower temperatures. For example, the Diels-Alder reaction is carried out at 0 °C. Also, during the generation of diazomethane, the temperature is kept below 0 °C to prevent the decomposition

Appendix A

of diazomethane. The following methods are used for cooling:

A. **Ice-Baths:** A mixture of crushed ice and water in a beaker should be kept handy during violent reactions (e.g., Grignard) to slow the rate of reflux, or placed beneath a receiving flask for the collection of low boiling products. The temperature of such an ice-bath is generally between 3° and 5°.

B. **Salt Added to Ice-Baths:** Regular salt (sodium chloride) can be added to ice-baths to lower the temperature to about -10°. Such ice-baths are used to induce crystallization of certain compounds from solutions.

DRYING PROCEDURES FOR LIQUIDS AND SOLVENTS

Liquids and solvents containing reaction products that have come into contact with water are said to be wet. During the work-up/purification process, these wet products need to be dried. This is accomplished by a process known as chemical drying. A solid drying agent is added to the liquid which reacts with the water present and forms a hydrated compound.

$$\text{compound in hexane} + n\text{H}_2\text{O} \xrightarrow{\text{MgSO}_4} \text{compound} + \text{MgSO}_4 \cdot n\text{H}_2\text{O}$$

The drying agent should not react chemically with the liquid or form emulsions.

The following desiccants are recommended if one is not specifically mentioned in the work-up procedure of the experiment:

1. $CaSO_4$ forms a dihydrate, $CaSO_4 \cdot 2H_2O$, is efficient and rapid, but has a low capacity.
2. $MgSO_4$ forms multiple hydrates, $MgSO_4 \cdot nH_2O$, is an excellent drying agent, but requires careful filtration for removal once it is hydrated.
3. Na_2SO_4 forms a decahydrate, $Na_2SO_4 \cdot 10H_2O$, has a high capacity, but requires more time to accomplish the drying.

SOLIDS

A solid can generally be dried by spreading it on a piece of filter paper and then storing it in a sample vial covered with foil which is perforated in order to allow air circulation. The vial can be placed in a beaker and left overnight in a drawer. It can also be placed in a small beaker and left in a drying oven with care taken that the temperature of the oven is well below the melting point of the compound. A third method is to use a vacuum dessicator.

For research grade drying, solids are heated under vacuum in the Abderhalden drying apparatus (also known as a "drying pistol").

Appendix A

RECORDING A MELTING POINT (RANGE)

MELTING POINT

The melting point (mp) of a pure compound is defined as that temperature at which the solid and liquid phases of the compound are in equilibrium at a pressure of one atmosphere (760 mmHg of mercury). The melting point is probably the single most important criterion for determining the purity of a solid compound. The melting point of a pure compound is generally quite sharp, usually melting within a degree or two, whereas an impure compound will melt over a wider range. As a rule of thumb, if the melting point range of a particular compound exceeds 3 °C, then the compound is impure and requires purification.

Mixed Melting Point:
Since very few compounds have the same melting point, this physical property is also used for the determination of the identity of organic compounds.

1. The identity of a particular compound can be verified further by carrying out a mixing melting point determination. When a pure sample of the suspected compound is mixed with your unknown compound, and the melting point remains unaltered, the two compounds forming the mixture are declared identical. This is because the melting point of a mixture of two different compounds is usually lower than the melting points of either of the individual compounds.
2. Preparation of a Sample for a Melting Point Determination: A small amount (5-10 mg) of the solid is placed on a clean piece of filter paper and is crushed to a fine powder by the application of gentle pressure with a clean metal spatula and scraped into a small mound. The finely powdered solid is packed densely to a height of 3-5 mm in a melting point capillary tube (a capillary tube closed at one end). This is done by filling one of the melting point tubes by pushing the open end into the mound of powder, using the spatula as a backstop. When a small plug of powder has collected in the opening of the capillary tube, work the material down to the sealed end by scratching the capillary with a file while holding it lightly at the top. Repeat this process until a column of powder about 3 - 5 mm in height has collected in the capillary tube. Tap the powder compactly in the capillary, as for example, by dropping it on the desk several times through a 2-foot length of glass tubing.
3. Melting the Sample: Insert the melting-point tube in one of the three capillary wells of the Mel- Temp apparatus (see illustration). Rotate the voltage control knob to "0" and turn on the apparatus. Make certain that the sample in the melting-point tube is visible in the viewing lens. Adjust the voltage control knob to a setting that will give a reasonably rapid rise (about 5 $°C/_{minute}$) of temperature to within about 15 - 20 °C below the anticipated melting point. (To determine the voltage setting to be employed, use the family of curves in). When this temperature has been reached, quickly lower the voltage setting to a value that will give a heating rate of about 2°C per minute during the actual determination of the melting point.

You should be able to get satisfactory results by using an initial voltage setting of 60 volts, then at a temperature of 15°C - 20°C below the anticipated melting point quickly lowering the voltage to 35 volts if your sample should melt between 100°C and 120°C, to 40 volts if it should melt between 120°C and 140°C, or to 45 volts if it should melt between 140°C and 160°C.

4. Recording the Melting Range: Record the temperature at which you see the first sign of liquid and record the temperature at which the solid-liquid transition is complete. This is the melting range of the sample.

5. Do not use the capillary tube and sample for a second melting point determination. If you need to recheck the melting point recorded on your sample, you need to repeat the procedure previously described. It is generally necessary to wait for a time between melting point determinations in order to allow the apparatus to cool. Inaccurate results may be obtained if the waiting period is neglected. It is also a good idea to use the same apparatus for all the melting point determinations to be run during a single experiment. Be sure to turn the voltage control to "0" and turn off the apparatus as soon as you have finished.

Cause of Incorrect Melting Points.

1. It is important that the sample be dry and thoroughly crushed before it is introduced into the capillary tube.
2. The rate of heating is critical—too rapid a rate of heating may cause a deviation in the melting range from the expected values.
3. Impurities generally lower the melting range.

HOW TO PACK PRODUCT IN CAPILLARY TUBE (METHOD 2)

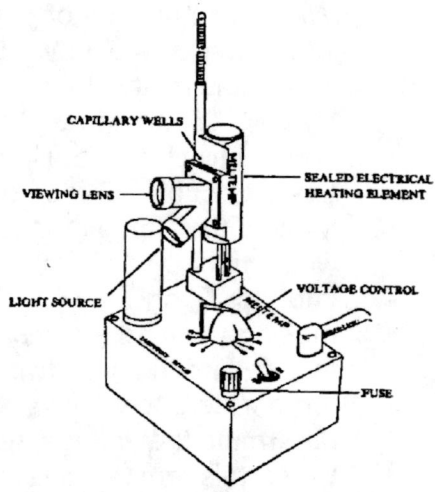

1. Pack the capillary tube by pressing the open end gently into a sample of the compound to be analyzed. Crystals will stick in the open end of the tube.
2. The solid should fill the tube to a depth of 2-3 mm.
3. Tap the bottom of the capillary on a hard surface so that the crystals pack down into the bottom of the tube. Alternatively, drop the capillary tube down a length of glass tubing or a glass **condensing column** to pack the crystals into the bottom of the tube.
4. When the crystals are packed into the bottom of the tube, place the tube into the slot behind the eyepiece on the Mel-Temp. Make sure the unit is plugged in and set to zero, and then turn it on. When

finished, turn off both the unit and the thermometer and place the used melting point tube in the used melting point capillary tube receptacle.

GRAVITY FILTRATION

This method involves the filtration of a solution through a paper filter held in a funnel, allowing gravity to draw the liquid through the paper. Gravity filtration is used to free a solution from solid impurities. The solid collected in the paper is discarded while the filtrate is saved. A gravity filtration can be either "hot" or "cold".

Regular (cold) gravity filtration requires a long stem funnel. It is usually accomplished by supporting the funnel on an Erlenmeyer flask. It is advisable to clamp the neck of the flask via an extension clamp to a ring stand.

When performing a hot gravity filtration (such as in Recrystallization) the following procedure should be followed:
1. Always use a short, wide stem funnel. With these types of funnels, it is less likely that the stem of the funnel will become clogged due to the accumulation of solid material, which is very likely to occur when a hot saturated solution comes into contact with a relatively cold funnel (or a cold flask) whereupon the solution will become cooled, supersaturated and crystallization will begin. Crystals will form in the filter and either fail to pass through the filter paper or clog the stem of the funnel.
2. Keep the solution to be filtered as near its boiling point as possible.
3. Preheat the funnel by pouring hot solvent through it prior to the actual filtration in order to minimize cooling of the solution as it strikes the apparatus.
4. Keep the filtrate (filtered solution) in the receiving container hot enough to continue boiling slightly (e.g. by setting it on a steam bath). The boiling solvent and the steam heat the receiving flask and the funnel stem and keep them clear of crystals.
5. Always support the short stem funnel using a ring support clamped on to a ring stand with an Erlenmeyer flask or a beaker placed just below the stem funnel. Do not support the funnel on the flask; doing so will slow down the filtration process and lead to crystallization on the filter paper.
6. It is useful to accelerate the filtration process by using fluted filter paper. The procedure for making it is illustrated below.

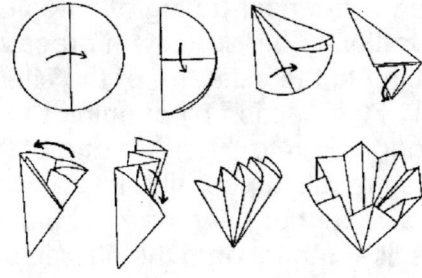

Appendix A

VACUUM SUCTION FILTRATION

Vacuum or suction filtration is more rapid than gravity filtration. It is used when an organic solid product needs to be collected from a primarily liquid medium. The solid is saved, washed and air dried, while the filtrate mayor may not be necessary. In many (but not all) cases, the filtrate can be discarded. The procedure for setting up a vacuum filtration follows:

1. A receiver flask with a side arm (a filter flask) is used. The side arm is connected by means of heavy-walled rubber tubing to a source of vacuum. A Buchner funnel is sealed to the filter flask by the use of a rubber stopper or a rubber gasket.
2. The flat bottom of the Buchner funnel is covered with an unfolded piece of circular filter paper which is held in place by suction.
3. It is advisable to moisten the paper with a small amount of solvent before beginning the actual filtration. The moistened filter paper adheres more strongly to the bottom of the Buchner funnel.
4. Under vacuum, a solution poured into the Buchner funnel is literally "sucked" through the filter paper at a rapid rate.
5. The most common source of vacuum in the laboratory is the water aspirator (approx. 10-20 mm Hg). This device passes water rapidly past a small hole to which a side arm is attached. The Bernoulli Effect gives rise to a reduced pressure along the side of the rapidly moving water stream and creates a partial vacuum in the side arm.

   Note:
   A. The filter paper should fit exactly in the center of the funnel. For instance, you would use a piece of filter paper having a diameter of 4.25 cm with the filter funnel having the same diameter. Failure to do so will result in loss of product. Under no circumstances should you attempt to cut filter paper to fit the funnel.
   B. The filter flask serves to collect the filtrate (the mother liquor). The side arm serves as a means of connecting the aspirator via pressure tubing (thick walled rubber tubing) to the assembly.
   C. It is important that thick walled rubber tubing be used in this procedure. Thin walled condenser tubing often collapses under vacuum and causes filtration to be slow. As a precaution, before connecting the rubber tubing to the side arm of the filter flask, check the efficiency (intensity) of the suction by opening the aspirator completely and placing your thumb to the other end of the pressure tubing. A weak suction may result in poor filtration and cause long delays.
   D. A correct size rubber bung or a flat piece of rubber may be used to attach the filter funnel onto the filter flask. This connection must be tight to ensure maximum vacuum.

E. A vacuum filtration set-up must be clamped via an extension clamp or a suitable three-finger clamp to a ring stand or support. If it is not, a mistaken swipe of the arm can easily ruin the assembly.

RECRYSTALLIZATION

Because they both involve the disappearance of a solid, the processes of melting and dissolving are sometimes confused. Melting is the simpler process: it is the changing of a substance from the solid to the liquid state.

$$\text{solid} \rightarrow \text{liquid}$$

The process of dissolving (or dissolution) involves two substances: one, usually a solid, is termed the solute while the other, usually a liquid, is termed the solvent. In the process of dissolving, the two substances become mixed into a single phase, usually a liquid, known as a solution.

$$\text{solute} + \text{solvent} \rightarrow \text{solution}$$

A familiar example is the dissolution of sugar (the solute) in water (the solvent) to give a sugar solution. The reverse process is known as precipitation or crystallization.

$$\text{solution} \rightarrow \text{solute} + \text{solvent (or a less concentrated solution)}$$

A simple way to purify a solid is to select a solvent that will dissolve it but in which the impurity is insoluble. The impurity is removed by filtration; the solution that comes through the filter (the filtrate), then yields the purified compound on evaporation of the solvent. Though this technique of purification by filtration is adequate in some instances, often some of the impurities also dissolve in the solvent. The purification method known as recrystallization can be used in these cases.

An organic solid may have insoluble and/or soluble components as impurities.

Recrystallization is a simple yet efficient way of separating these impurities from the pure compound. In its simplest form, recrystallization of a solid involves selecting a solvent that will dissolve the solid only upon heating the mixture. The solubility of the solid (solute) in a solvent depends on the temperature. Most organic compounds are more soluble in a hot solvent than in the same solvent at a lower temperature. This results in the formation of hot saturated solution with the insoluble impurities remaining in suspension. The insoluble impurities may be removed at this point by performing a hot gravity filtration.

The resulting solution is called the filtrate. It is a saturated solution of the solute in the solvent. If the solid was contaminated by some soluble impurity, this foreign substance will usually remain dissolved in the mother-liquor (filtrate).

The next step involves cooling the filtrate. A solid that saturates a solvent at an elevated temperature will crystallize out of solution on cooling. The crystallized compound is then separated from the mother-liquor and the soluble impurity by doing a vacuum filtration. The yield of the desired compound and its purity are improved if washing steps are included in the procedure.

A good solvent for use in this combined purification technique should have the following characteristics:
  a. It should not react with the solute;
  b. It should dissolve considerably more of the solute when hot than when cold;
  c. All impurities present should either be insoluble in the hot solvent or be soluble in the cool solvent; and
  d. The solvent should evaporate readily from the recrystallized solute.

SIMPLE DISTILLATION

When a pure liquid is distilled, vapor rises from the distilling flask and comes in contact with a thermometer. The vapor then passes through a condenser which condenses the vapor into droplets and passes it into the receiving flask. The temperature observed during the distillation of a pure substance will remain constant throughout the distillation as long as both vapor and liquid are present in the system (see Part A in the figure below). When a liquid mixture is distilled, often the temperature will not remain constant, but rather it will increase throughout the distillation. This occurs because the composition of the vapor which is distilling varies continuously during the distillation (see Part B in the figure below).

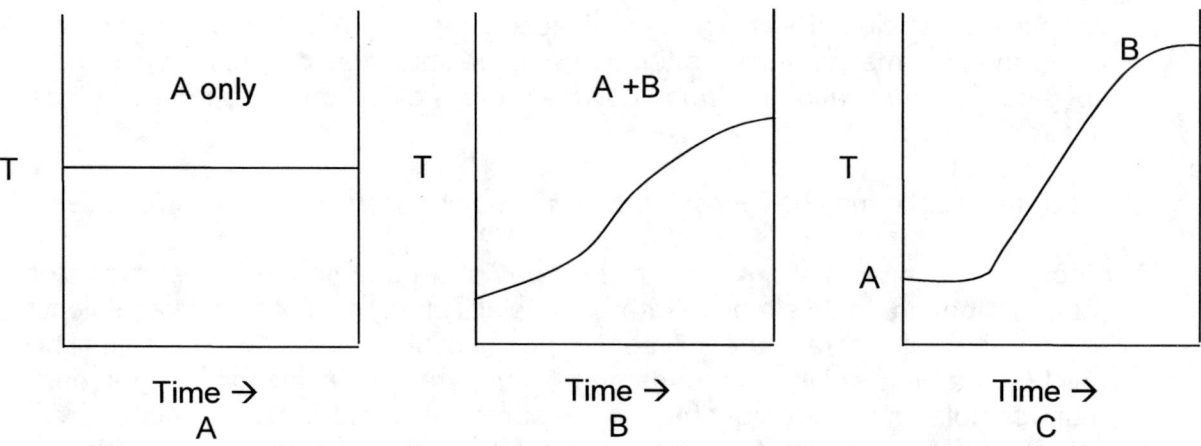

Three types of temperature behavior during a simple distillation: A, a relatively pure component is being distilled; B, a mixture of two components of similar boiling point is being distilled; C, a mixture of two components with widely differing boiling points is being distilled. Good separations are achieved in A and C.

Appendix A

When two components which have a large boiling point difference are distilled, one will observe that the temperature remains constant while the first component distills. If the temperature remains constant, a relatively pure substance is being distilled. After the first substance distills, the temperature of the vapors will rise, and then the second component will distill, again at a constant temperature. This is shown in the above figure, part C. A typical application of this type of distillation might be an instance of a reaction mixture containing the desired component A (boiling point 140°C) contaminated with a small amount of undesired component B (boiling point 250°C) and mixed with a solvent such as diethyl ether (boiling point 36 °C). The ether is easily removed at low temperature. Pure A is removed at a higher temperature and collected in a separate receiver. Component B could then be distilled, but it is usually left as a residue and not distilled. This separation would not be difficult and would represent a case where simple distillation might be used to advantage.

The procedure for assembling a simple distillation apparatus follows:

1. Place a few boiling aids in a round bottom flask (the size of the round bottom flask depends on the volume that needs to be distilled. As a general rule, one should use a round bottom flask which will hold twice the volume of the liquid to be distilled).
2. Attach a suitable size three-finger clamp (or an extension clamp) at the neck of the round bottom flask and attach this securely approximately halfway up a ring stand.
3. Apply grease very sparingly to the upper half of the two ground glass joints of the side-arm connector. Slide the greased lower vertical joint into the neck of the round bottom flask. Rotate the side arm connector gently until a thin and even film of grease is spread over the entire connection. This assembly is Unit A.
4. Secure two, four-foot lengths of thin walled condenser tubing to the water inlet and outlet extensions of the distilling condenser. The condenser tubing should be attached securely such that it almost makes contact with the outer surface of the distilling column. It is easier to slide the tubing on if the end of the condenser tubing is dipped in cold water.
6. Attach a 3-finger clamp (or an extension clamp) to the center of the condenser and secure this lightly to a ring stand. This assembly is Unit B.
7. Bring the two ring stands together and adjust the height of B so that Unit A and Unit B may be connected via the side arm connector.
8. Release the clamp supporting the condenser slightly and rotate the condenser in order to spread a thin film of grease over the entire connection. This operation must be done carefully, making sure that a firm and secure connection is made between the two units.
9. Once the two units, A and B, are well connected, tighten clamps and supports securely.

Appendix A

10. Lightly grease the upper end of the terminal (free) ground glass joint of the distilling column and attach the vacuum adapter using a stirrup clamp (if available) for support. Alternately, rubber bands may be used to support the vacuum adapter.
11. Place the receiver (10 mL graduated cylinder, a small beaker, etc.) below the vacuum adapter.
12. Introduce the liquid into the distilling flask through the open end of the side arm connector using a long stem funnel. This operation must be done carefully in order to prevent the liquid from entering the side arm of the distilling head (side arm connector).
13. Apply a smear of grease on the bulb of the thermometer. Slide the thermometer into its rubber adapter using a gentle circular motion. A thermometer must never be forced into its adapter.
14. Wipe off excess grease on the bulb of the thermometer with a soft tissue moistened with hexane.
15. Connect the thermometer with its rubber adapter to the rest of the assembly by means of a ground glass thermometer adapter.
16. The heating mantle should be attached to the ring stand immediately below the round bottom flask independent of the glass assembly. The round bottom flask should not be supported by the heat source. At any given time during the distillation it should be possible to turn off the heat supply quickly by lowering the heating mantle, rather than by turning off the power supply, which would take a while to cool the distillation flask. The heating mantle should not be connected directly to the main outlet. It should be plugged into the Variac heat control unit located on the bench. Never use a heat source to support an assembly of glassware.
17. As a routine procedure, it is advisable to wrap the side-arm connector with a layer of glass wool. This serves to insulate the sidearm connector from air currents, which tend to have a cooling effect, thus causing the liquid to reflux and lengthening the distillation. This is found to be particularly useful when distilling high boiling liquids (boiling point > 125°).
    a. The glass wool should be wrapped such that periodic examination of the temperature is possible.
    b. Under no circumstances should you use paper towels or Kimwipes to insulate the assembly.
    c. Glass wool is an eye (and skin) irritant.
    d. Be cautious when handling it and wear safety glasses.
    e. After the experiment is completed, remove glass wool from the assembly and place it in the proper container.
    f. Under no circumstances should glass wool be left in your work area.
18. After the experiment is over, dismantle the set-up in exactly the reverse order, and do it immediately after use to prevent the joints from sticking.
19. A convenient way of cleaning glassware that has been greased is to wipe the ground glass joints with Kimwipes, followed by wiping with Kimwipes moistened in hexane.

Appendix A

THE FOLLOWING OBSERVATIONS APPLY TO SIMPLE DISTILLATION:

a. The thermometer should extend slightly below the side arm leading to the condenser.
b. The cooling water should come in from the lower end of the condenser and exit from the top.
c. Ensure that all connections are tight.
d. The reaction flask (distilling flask) should not be more than half full.
e. Boiling aids must be added to the flask to prevent bumping. Do not add boiling aids to a hot solution since this will cause the liquid to erupt violently.
f. Wrap the distillation set-up with glass wool. This should be done especially if there is any difficulty observed during distillation, e.g., with a high boiling liquid.
g. If a low boiling-solvent is distilled, the receiving flask should be immersed in an ice-bath to minimize volatilization of the solvent.
h. Sometimes a distilling liquid may solidify during its passage through the condenser. This is observed in the case of a low melting solid, e.g., tertiary butanol (especially true during the winter months when the water flowing through the condenser is quite cold). If this phenomenon is observed, merely discontinue the water supply to the condenser and observe for 5-10 minutes. If the situation still persists, obtain a heat gun and, under your instructor's supervision, gently warm the distilling condenser with a current of hot air.
i. Do not heat any solution to dryness; this may lead to contamination of the distillate by decomposition products, or sometimes, if peroxides are present, may cause an explosion.

Micro-Boiling Point Determination:

Sometimes when the amount of liquid obtained in your final product is a small amount (>3mL) you can use a micro-boiling point to determine the boiling point of your liquid.

1. Take your sample and pour it into a clean dry small test tube.
2. Obtain a small melting point capillary tube from the lab supply drawer and place it into the liquid filled test tube closed side up.
3. Take this test tube and secure the thermometer with a rubber band so that the bottom of your test tube is the same as your thermometer bulb.
4. Place the entire assembly into an extra large test tube filled with mineral oil so that the entire level of the liquid in the test tube is below that of the mineral oil 'bath'.
5. Heat the mineral oil bath using your heating mantle until you see bubbles of gas begin the escape the bottom of the capillary tube.
6. Once you see a steady stream of bubbles coming off the bottom end of your inverted melting point tube remove the extra large test tube from the heat source and allow the apparatus to cool.
7. As the liquid cools the atmospheric pressure will equal the vapour pressure of the liquid and the liquid will begin to rise up the capillary tube. Record your temperature - this is your boiling point.

A SIMPLE REFLUX APPARATUS

A reaction done under reflux suggests heating a reaction mixture containing liquid or solid reactants with solvent (an inert medium for the reaction) without the loss of volatile solvent or liquid. When the temperature of the reaction mixture reaches the boiling point of the solvent, the mixture starts to boil, and the solvent turns into a vapor. A reflux condenser provides a cold surface. The hot solvent vapor, on contact with the cold inner surface of the condenser, condenses into a liquid which then drops back into the reaction vessel. The liquid level therefore remains virtually constant.

1. Use a round bottom flask which will hold more than twice the volume of the reactants.
2. Clamp the neck of the round bottom flask with an appropriate size three-finger clamp (or extension clamp) and secure this firmly to a ring stand.
3. Place the reactants in the round bottom flask. A solvent or liquid reactant should be poured carefully into the flask preferably using a long stem funnel. A solid reactant should be introduced using a powder funnel.
4. Place a few boiling aids in the round bottom flask.
5. Attach two, four foot pieces of thin walled rubber tubing to the outlet and inlet of the West condenser. Make sure that these connections are tight.
6. Apply a very small amount of grease around the top half of the ground glass joint of the condenser that fits into the neck of the round bottom flask.
7. Insert the condenser into the round bottom flask. Rotate the condenser gently until a thin and even film of grease is spread over the entire connection.
8. A suitable sized heating mantle is placed under the round bottom flask. The heat source should be arranged and secured independent of the rest of the glassware. The glassware should, under no circumstances be supported by the heat source. Occasionally during a reflux, the reaction may become too vigorous. It is more convenient to be able to reduce the heat constant of the system quickly by lowering or removing the heat source, thus cooling is effected almost immediately. The alternate methods of turning the power supply off or lowering the setting require a longer period of time to achieve the same results.
9. If a steam bath is used as a source of heat, the assembly should be clamped in such a manner that the glassware can be lifted up quickly away from the heat if the need arises during an experiment. For efficient heating, a heating mantle or a steam bath should fit snugly below the reaction flask. Whenever possible, use the correct size heating mantle.
10. The rubber tubing closest to the round bottom flask (inlet) is connected to the faucet. The upper tube (outlet) is placed in the sink.
11. The faucet should be opened cautiously and the flow of water through the condenser should be moderate.

Appendix A

USE OF THE SEPARATORY FUNNEL

The transfer of a substance from one liquid to another is called liquid-liquid extraction. Extractions are widely used and are relatively simple processes. Many organic reactions are followed by work-ups employing some kind of extraction as part of the purification procedure.

Liquid-liquid extractions are performed with a separatory funnel which has been checked to see that the stopper and stopcock fit properly. Glass stopcocks should be greased slightly with stopcock grease; Teflon stopcocks do not need grease.

Proper Procedure for Handling the Separatory Funnel:

1. With the stopcock closed, support the funnel in a ring on a ring stand.
2. Add the solution to be extracted to the funnel, and then add the extraction solvent. It is essential that the solution be cool so that low boiling solvents will not volatilize, create pressure within the funnel, and blow out the stopper with hot or dangerous contents.
3. Wet the stopper with water to keep it from sticking tightly and so that organic solvents are less likely to leak out.
4. The extracting solvent is usually used in a volume of about one-fourth to one-third the volume of the solution containing the product.
5. Hold the funnel with both hands- one hand on the stopcock and stem, and one hand on the stopper and neck. Rock the funnel gently back and forth lengthwise once.
6. Point the stem away from your face and open the stopcock to release any excess vapor pressure. Repeat this procedure until only a small spit of pressure is observed. (More vigorous mixing can be entertained, but also with occasional opening of the stopcock to test for pressure.)
7. After mixing, hold the funnel upright, and swirl it once to set up a gentle rotary motion of liquid inside the funnel. Then set it in the ring of the ring stand. (The swirl helps prevent the formation of droplets of one immiscible phase within another along the funnel walls.)
8. After allowing the two immiscible liquids to clearly separate, remove the stopper in order to admit air and allow the lower liquid to smoothly drain into an appropriate container. Drain the lower liquid to the point that the upper liquid barely reaches the stopcock hole.
9. Extractions are repeated with fresh solvent. It is unnecessary to remove the upper phase in the funnel when the upper phase contains the substance of interest, although it is important that none of the upper layer be allowed to drain in that case. Most organic liquids are less dense than water, but some, notably halogenated solvents will be more dense (e.g. chloroform, methylene chloride).
10. Whenever directions say to "wash" a solution, it means that you should extract the solution to remove impurities. Washing includes all of the steps in the extraction procedure.

Appendix A

11. Emulsions are a suspension of small droplets of one immiscible liquid in another. Commonly emulsions are troublesome and the two phases in the separatory funnel will not separate well. You may observe a mixture of droplets between two clear phases. An untrained observer might interpret the mixture as a third phase. Emulsions sometimes break up if allowed to stand for 5 – 10 minutes. If no improvement is noted, consult your TA. When emulsions are likely to form, the slow rocking motion must be used throughout the procedure.

APPENDIX B – GENERAL

Percent Yield

$$\frac{\text{Actual Mass of Product}}{\text{Theoretical Yield of Product}} \times 100\% = \text{Percent Yield}$$

Percent Recovery

$$\frac{\text{Mass Recovered}}{\text{Mass Used}} \times 100\% = \text{Percent Recovery}$$

Percent Difference

$$\frac{\left| T_{reference} - T_{sample} \right|}{T_{reference}} \times 100\% = \text{Percent Difference}$$

APPENDIX C – THEORETICAL YIELDS
CHEM 332L

Experiment 2: A Grignard Reaction

$$\frac{\_\_\_\_ \text{ g benzophenone}}{1} \times \frac{1 \text{ mol benzophenone}}{182.22 \text{ g benzophenone}} \times \frac{260.33 \text{ g triphenylmethanol}}{1 \text{ mol triphenylmethanol}} =$$

Experiment 3: An Aldol Condensation

$$\frac{\_\_\_\_ \text{ mL mixture}}{1} \times \frac{.044 \text{ mL acetone}}{1 \text{ mL mixture}} \times \frac{0.7857 \text{ g acetone}}{1 \text{ mL acetone}} \times \frac{1 \text{ mol acetone}}{58.08 \text{ g acetone}} \times \frac{234.30 \text{ g dibenzalacetone}}{1 \text{ mol dibenzalacetone}} =$$

Experiment 4: Sodium Borohydride Reduction

$$\frac{\_\_\_\_ \text{ g benzil}}{1} \times \frac{1 \text{ mol benzil}}{210.23 \text{ g benzil}} \times \frac{214.26 \text{ g hydrobenzoin}}{1 \text{ mol hydrobenzoin}} =$$

Experiment 6: Diels-Alder Reaction

$$\frac{\_\_\_\_ \text{ g maleic anhydride}}{1} \times \frac{1 \text{ mol maleic anhydride}}{98.06 \text{ g maleic anhydride}} \times \frac{274.27 \text{ g product}}{1 \text{ mol pruduct}} =$$

Experiment 7: EAS: Solvent-Free Claisen Condensation of Ethyl Phenylacetate

$$\frac{\_\_\_\_ \text{ mL ethyl phenylacetate}}{1} \times \frac{1.03 \text{ g}}{\text{mL}} \times \frac{1 \text{ mol}}{164.2 \text{ g}} \times \frac{1 \text{ mol ethyl}-3-\text{oxo}-2,4-\text{diphenylbutanoate}}{2 \text{ mol ethyl phenylacetate}} \times \frac{282.13 \text{ g}}{\text{mol}} =$$

Experiment 8: EAS: Nitration of Methyl Benzoate

$$\frac{\_\_\_\_ \text{ mL methyl benzoate}}{1} \times \frac{1.094 \text{ g methyl benzoate}}{1 \text{ mL methyl benzoate}} \times \frac{1 \text{ mol methyl benzoate}}{136.15 \text{ g methyl benzoate}} \times \frac{181.15 \text{ g methyl 3-nitrobenzoate}}{1 \text{ mol methyl 3-nitrobenzoate}} =$$

Experiment 9: EAS: Friedel-Crafts Alkylation

$$\frac{\_\_\_\_ \text{ g 1,4-dimethoxybenzene}}{1} \times \frac{1 \text{ mol 1,4-dimethoxybenzene}}{138.17 \text{ g 1,4-dimethoxybenzene}} \times \frac{250.38 \text{ g 1,4-di-t-butyl-2,4-dimethoxybenzene}}{1 \text{ mol 1,4-di-t-butyl-2,4-dimethoxybenzene}} =$$

APPENDIX D - SPECTROSCOPY

INFRARED SPECTROSCOPY

± 3400 cm^{-1} O—H of Alcohols, Phenols
               N—H of Amines, Amides

3200 – 2400 cm^{-1} O—H of Carboxylic Acids

B. ± 3000 cm^{-1} sp   C—H (3300 cm^{-1})
                     sp^2 C-H   (3100 cm^{-1})
                     sp^3 C—H (2950 / 2925 cm^{-1})

Aldehyde   (2800 cm^{-1})
            (2725 cm^{-1})

C. ± 1700 cm^{-1}
1. anhydride   1800 cm^{-1}
2. acid chlorides   1770 cm^{-1}
3. ester   1735 – 1740 cm^{-1}
4. aldehyde   1725 cm^{-1}
5. ketones   1710 cm^{-1}
6. carboxylic acids   1700 cm^{-1}
7. amides   1645 cm^{-1}

Conjugation will change the carbonyl stretch to lower frequencies in all of the above.

D. ± 1700 cm^{-1} alkenes (1650 cm^{-1})

aromatics   (1600 cm^{-1})
            (1475 cm^{-1})

E.   >C—C< 1100 cm^{-1} >C—O

F. –CH$_3$ bending vibrations: 1375 cm^{-1}
      –CH$_2$ bending vibrations: 1260 – 1280 cm^{-1}

1.) Fingerprint region: 1200 – 650 cm^{-1}
2.) Diagnostic region: 4000 – 1250 cm^{-1}

Appendix D

¹H-NMR SPECTROSCOPY

CORRELATION CHARTS

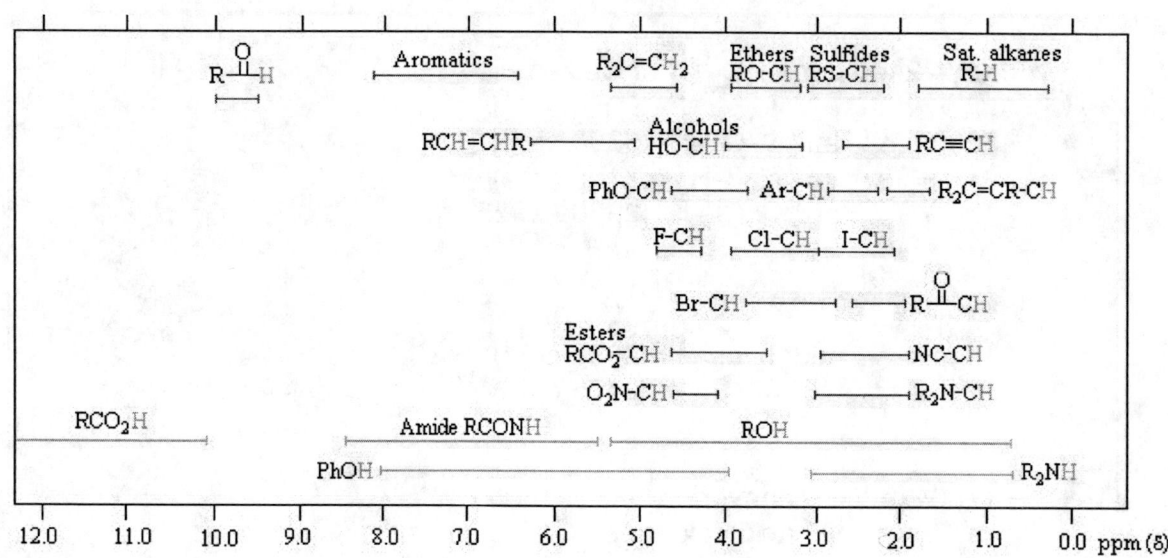

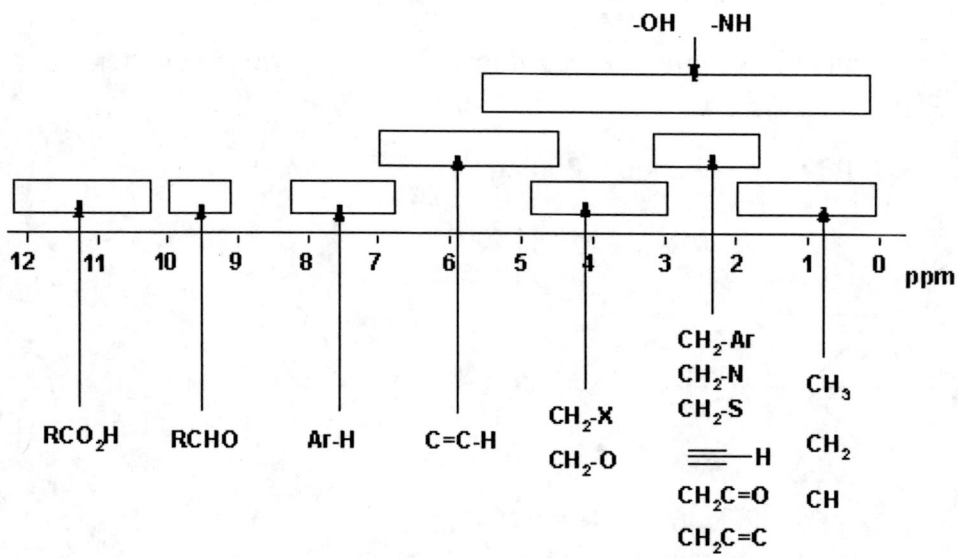

Appendix D

EXAMPLE PROBLEMS

For the following spectra the empirical formulas, integration, and splitting are given. Make the proper peak assignments and draw the compound.

1. C_2H_6O

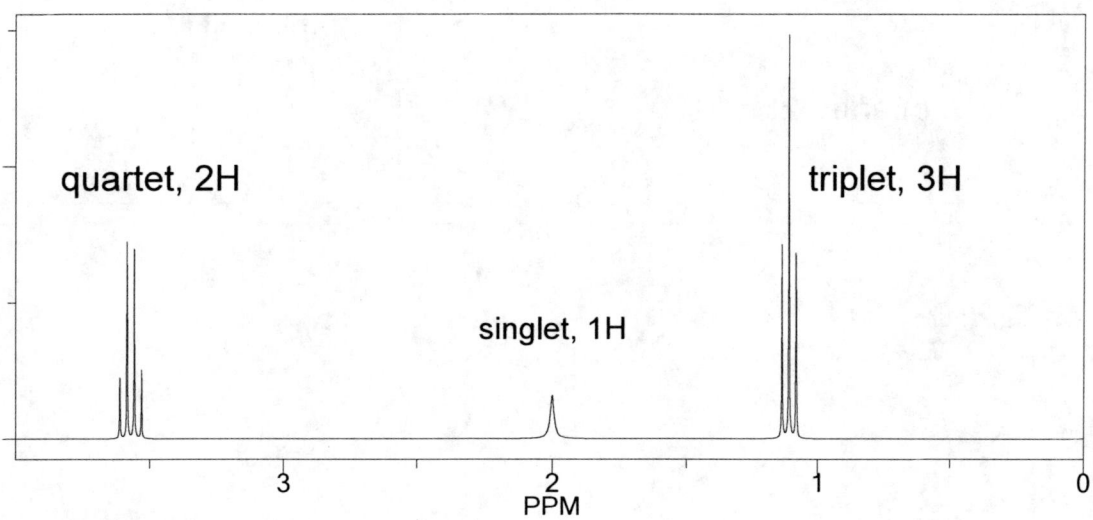

2. C_4H_9Br

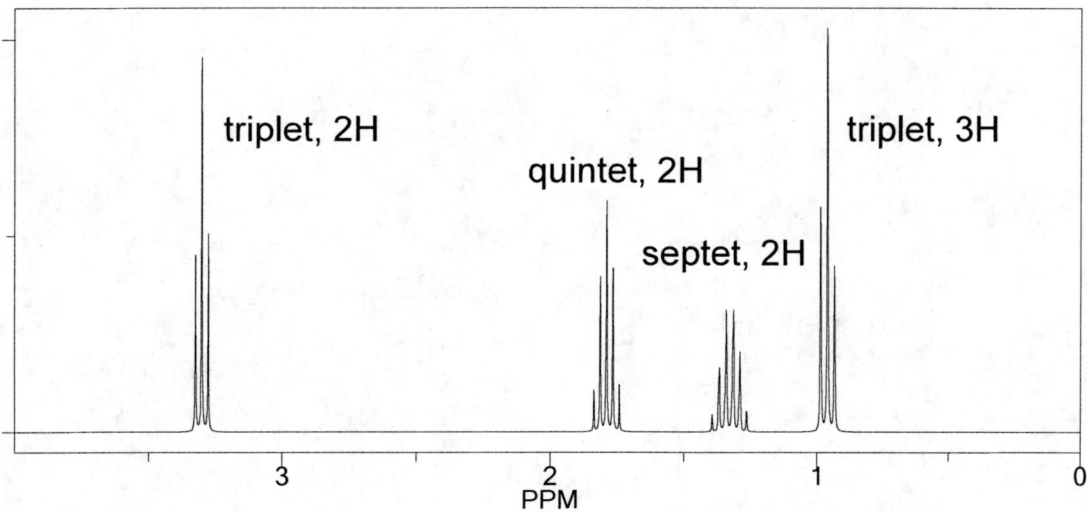

3. C₄H₈O₂

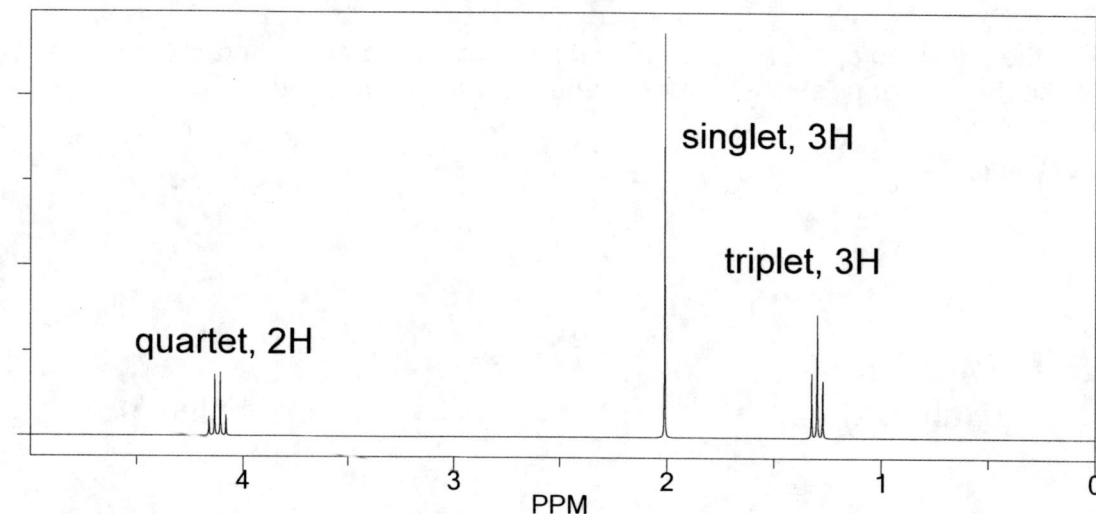

4. C₈H₈O₂

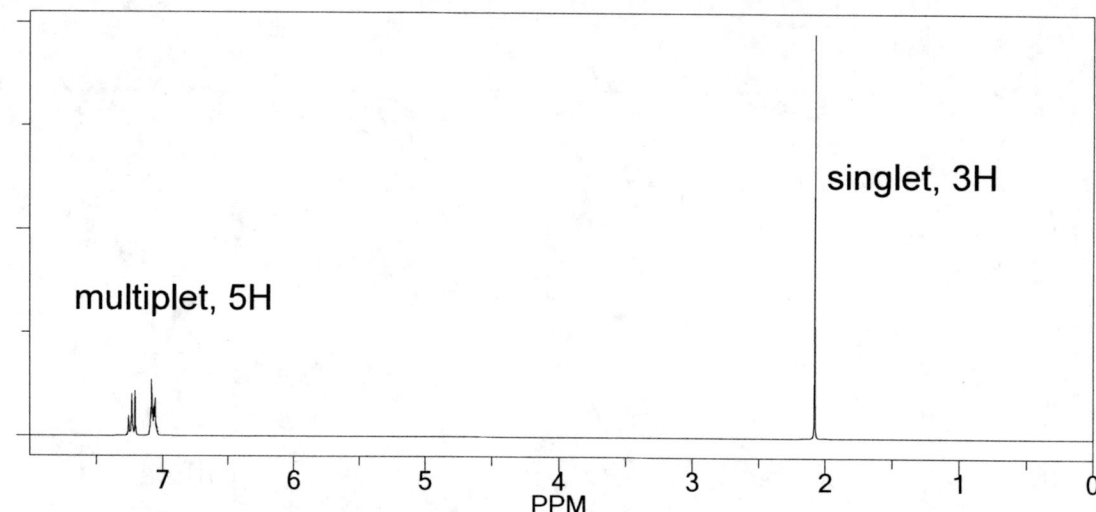

5. C₉H₁₀O₂

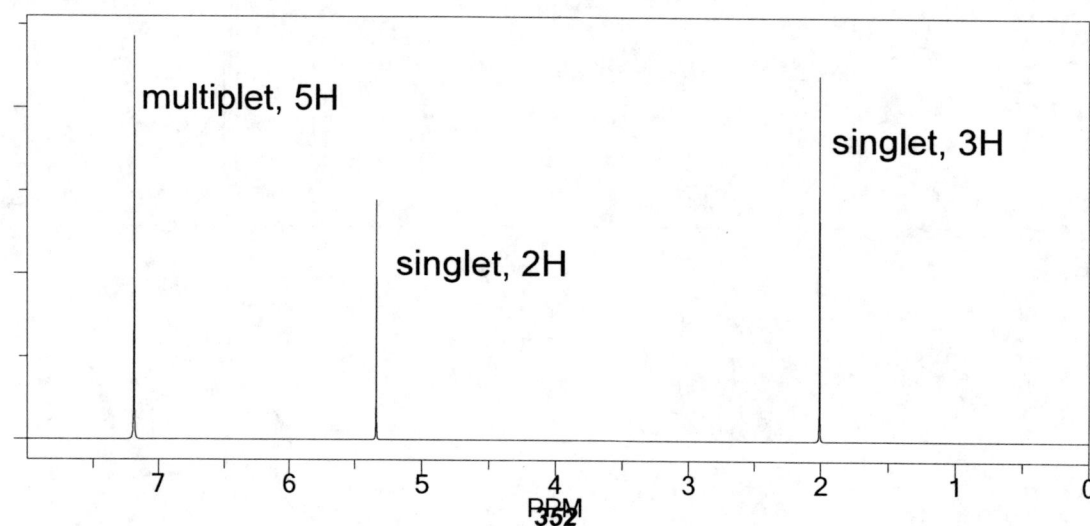

6. $C_4H_9NO_2$

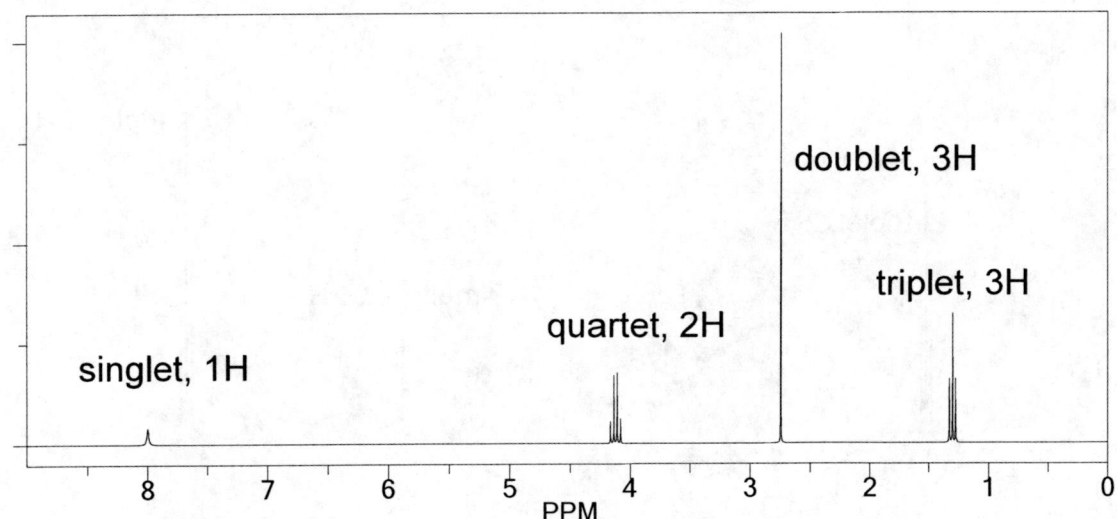

7. C_8H_8

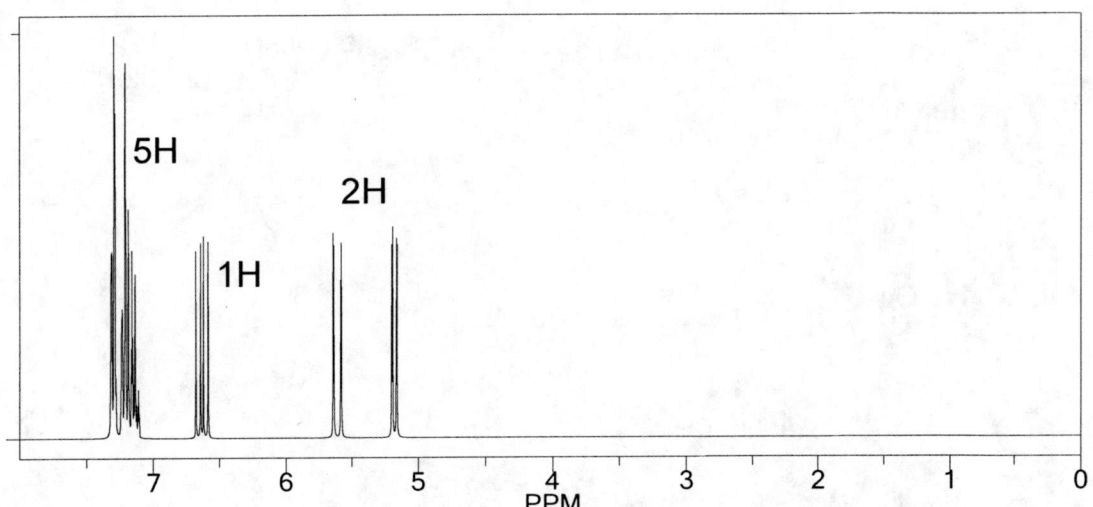

Can you explain the splitting patterns?

Appendix D

8. C₃H₇NO₂

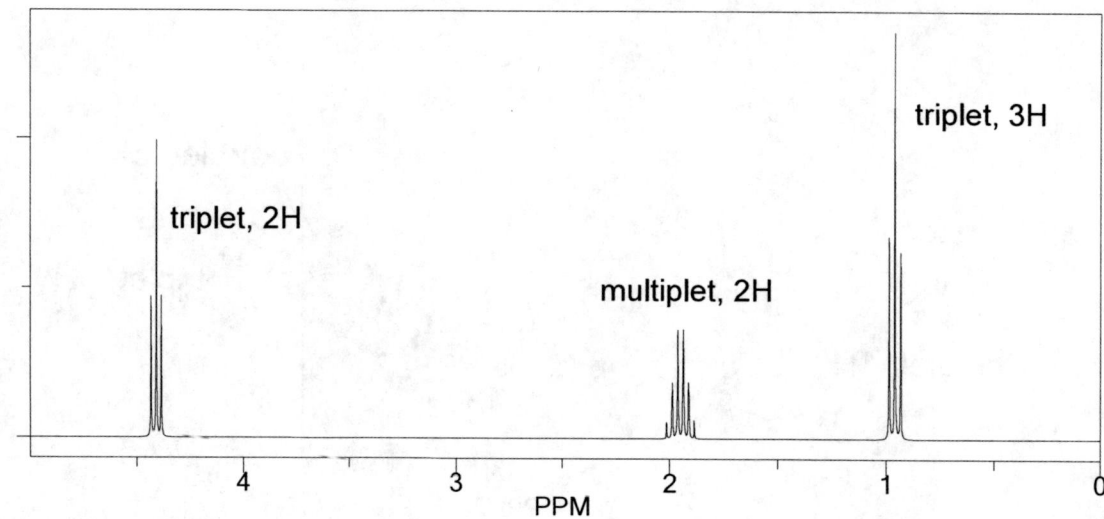

9. C₆H₁₀O₂

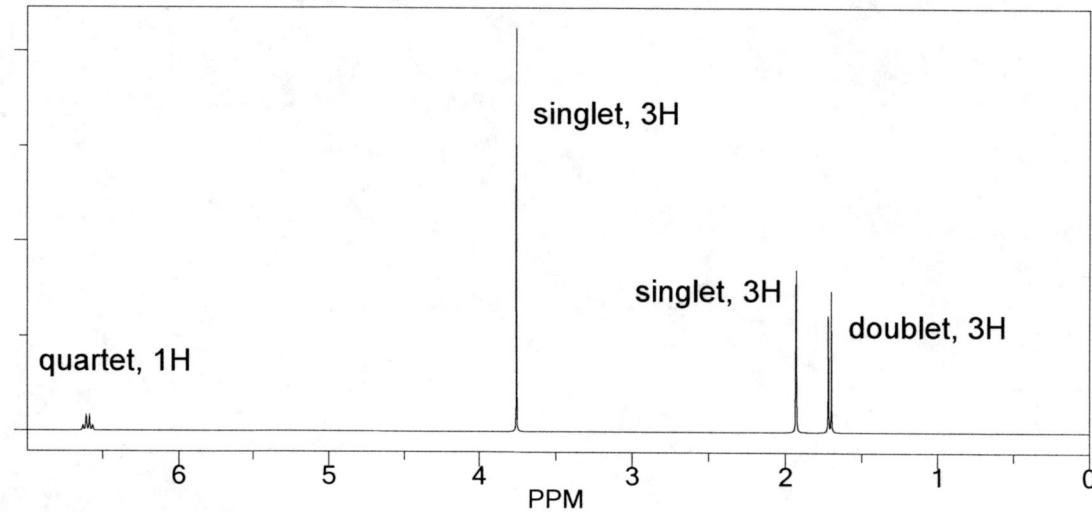

10. C₉H₁₆O₄

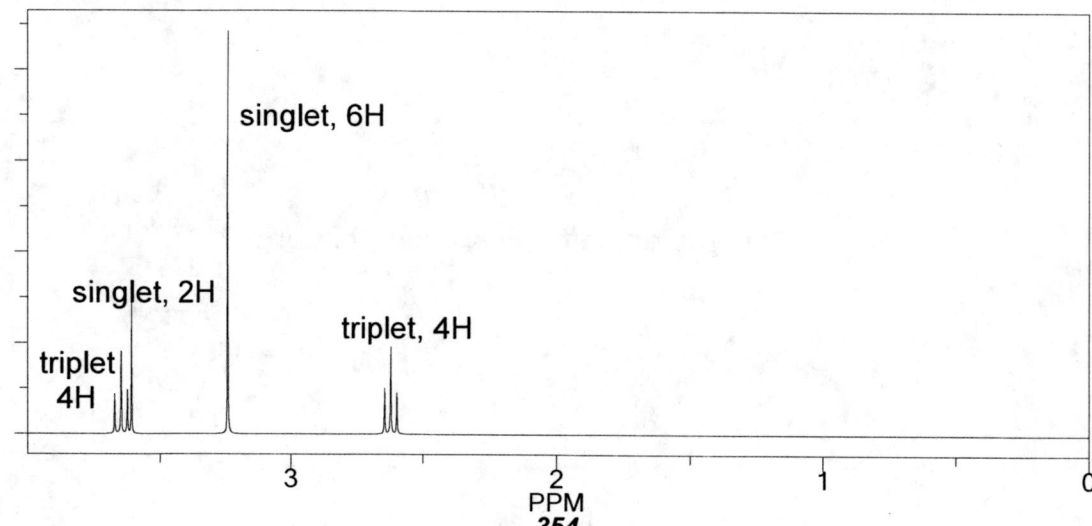

Appendix D

ANSWERS TO EXAMPLE PROBLEMS

1. C_2H_6O

   CH₃CH₂OH (ethanol)

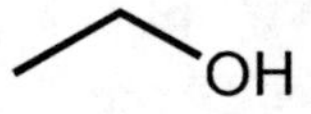

2. C_4H_9Br

   1-bromobutane

3. $C_4H_8O_2$

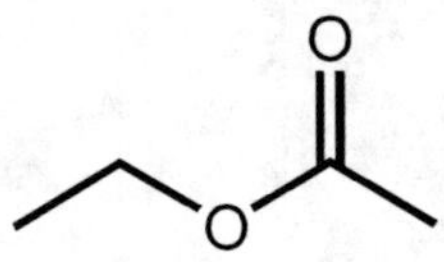

   ethyl acetate

4. $C_8H_8O_2$

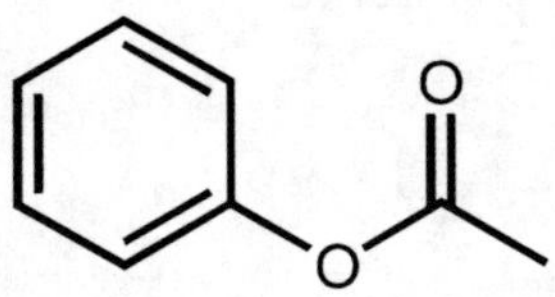

   phenyl acetate

5. $C_9H_{10}O_2$

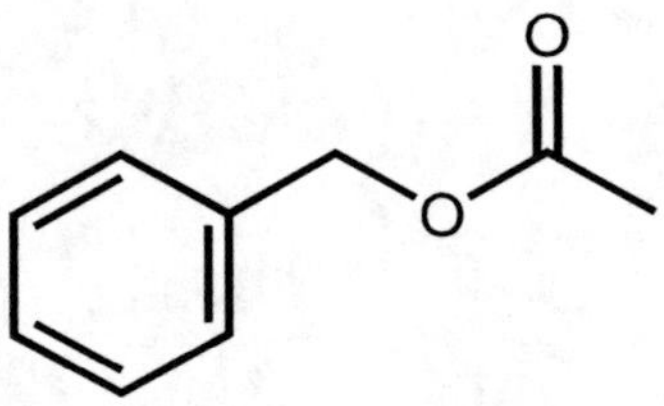

   benzyl acetate

6. $C_4H_9NO_2$

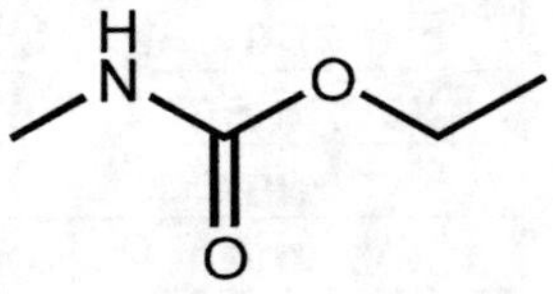

   ethyl N-methylcarbamate

7. C_8H_8

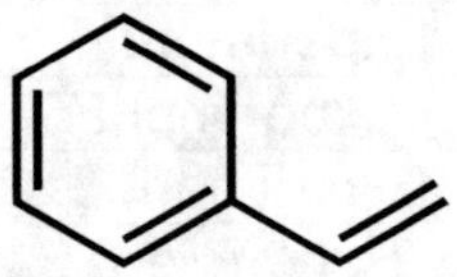

   styrene

8. $C_3H_7NO_2$

   1-nitropropane

9. $C_6H_{10}O_2$

   methyl (E)-2-methyl-2-butenoate

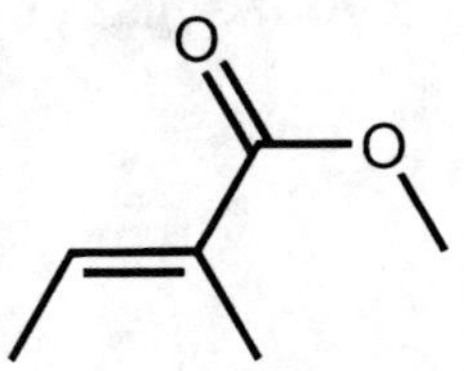

10. $C_9H_{16}O_4$

    1,7-dimethoxy-3,5-heptanedione

355

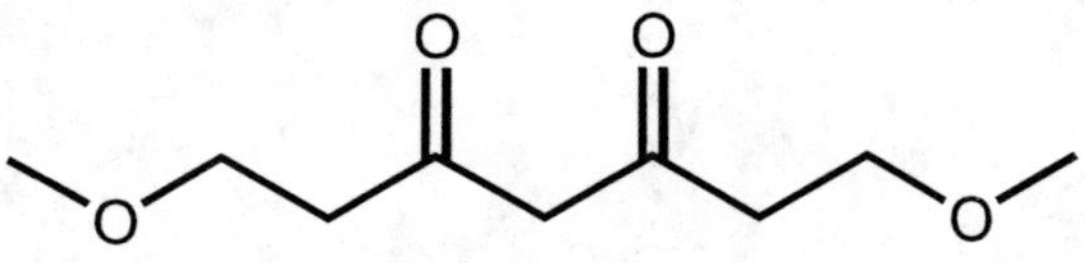

Appendix E

Grades CHEM 332L

Assignment	Grade	Points possible
Safety Quiz		100
Experiment 1, Quiz 1		50
Experiment 1, Report 1		50
Experiment 2, Quiz 2		50
Experiment 2, Report 2		50
Experiment 3, Quiz 3		50
Experiment 3, Report 3		50
Experiment 4, Quiz 4		50
Experiment 4, Report 4		50
Experiment 5, Quiz 5		50
Experiment 5, Report 5		50
Experiment 6, Quiz 6		50
Experiment 6, Report 6		50
Experiment 7, Quiz 7		50
Experiment 7, Report 7		50
Experiment 8, Quiz 8		50
Experiment 8, Report 8		50
NMR Worksheet		50
NMR Quiz		50
Final Exam		200
Total Points		**1200**

- These point assignments are subject to change at any time. Please consult your TA and your syllabus for current up to date points available.

EMERGENCY INFORMATION NOTES

NOTES